AF595121

2D Nanomaterials Processing and Integration in Miniaturized Devices

2D Nanomaterials Processing and Integration in Miniaturized Devices

Editors

Matteo Cocuzza
Fabrizio Pirri

MDPI • Basel • Beijing • Wuhan • Barcelona • Belgrade • Manchester • Tokyo • Cluj • Tianjin

Editors

Matteo Cocuzza
Department of Applied Science and Technology
Politecnico di Torino
Torino
Italy

Fabrizio Pirri
Department of Applied Science and Technology
Politecnico di Torino
Torino
Italy

Editorial Office
MDPI
St. Alban-Anlage 66
4052 Basel, Switzerland

This is a reprint of articles from the Special Issue published online in the open access journal *Micromachines* (ISSN 2072-666X) (available at: www.mdpi.com/journal/micromachines/special_issues/Nanomaterials_Processing_Integration).

For citation purposes, cite each article independently as indicated on the article page online and as indicated below:

LastName, A.A.; LastName, B.B.; LastName, C.C. Article Title. *Journal Name* **Year**, *Volume Number*, Page Range.

ISBN 978-3-0365-1570-0 (Hbk)
ISBN 978-3-0365-1569-4 (PDF)

Contents

About the Editors

Matteo Cocuzza

Matteo Cocuzza got his M.D. in Electronic Engineering from Politecnico di Torino, Turin, Italy, in 1997, and his Ph.D. degree in Electronic Devices in 2003. He is currently a Professor at the Dept. of Applied Science and Technology of Politecnico di Torino and associated researcher of IMEM-CNR. In 1998, he was one of the founders of the Chilab-Materials and Microsystems Laboratory. He is currently a lecturer of master's degree courses related to MEMS, micro- and nano-technologies, also in the framework of the International Master Degree in Nanotechnologies for ICT (joined master between Politecnico di Torino, INPG Grenoble and EPFL Lausanne). His research activity is focused on MEMS and microsensors for industrial applications, on the development of microfluidics and lab-on-a-chip for biomedical applications and, more recently, on the development and application of polymeric 3-D printing technologies.

Fabrizio Pirri

Candido Fabrizio Pirri is Full Professor at the Dept. of Applied Science and Technology of Politecnico di Torino and Responsible for the Chilab-Materials and Microsystems Laboratory since 1998. He joined the Center for Sustainable Future Technologies, Istituto Italiano di Tecnologia, Genova, Italy, as Director in 2011. Since 2005, he has been Director of the National MIUR Excellence Laboratory "Latemar", whose mission is the development of advanced technologies for genomics and proteomics. He is a lecturer of courses on physics of matter and introduction to nanotechnologies with Politecnico di Torino.

Editorial

Editorial for the Special Issue on 2D Nanomaterials Processing and Integration in Miniaturized Devices

Candido Fabrizio Pirri [1,2,*] and **Matteo Cocuzza** [1,3,*]

1 Department of Applied Science and Technology (DISAT), Politecnico di Torino, C.so Duca degli Abruzzi 24, 10129 Turin, Italy
2 Center for Sustainable Future Technologies, Istituto Italiano di Tecnologia, Via Livorno 60, 10144 Turin, Italy
3 Institute of Materials for Electronics and Magnetism, IMEM-CNR, Parco Area delle Scienze 37/A, 43124 Parma, Italy
* Correspondence: fabrizio.pirri@polito.it (C.F.P.); matteo.cocuzza@infm.polito.it (M.C.)

Citation: Pirri, C.F.; Cocuzza, M. Editorial for the Special Issue on 2D Nanomaterials Processing and Integration in Miniaturized Devices. *Micromachines* **2021**, *12*, 254. https://doi.org/10.3390/mi12030254

Received: 25 February 2021
Accepted: 1 March 2021
Published: 2 March 2021

Initially considered little more than a scientific curiosity, the family of 2D nanomaterials has become increasingly popular over the last decade. Starting from the undisputed progenitor, i.e., graphene, which to date has reached a technological maturity and a critical mass of knowledge to allow envisaging multiple applications in real manufacturing, numerous other materials have progressively been added to this family: transition metal dichalcogenides, boron nitride, metal oxide nanosheets, MXenes, layered double hydroxides, etc. Similarly, the range of applications under study and development has expanded, integrating and taking advantage of the unique properties of 2D nanostructured materials: sensors, optical applications, photovoltaics, touch screens, catalysis, filtration and exploitation as fillers to modulate the mechanical, electrical and chemical properties of the host matrices. Such a plethora of material/application combinations, in addition to the obvious technical requirements related to the synthesis of materials and the optimization of purity, composition, morphology and yield, also presents significant technological challenges for the corresponding processing, patterning and above all integration into systems and devices of higher dimensionality for the real exploitation of their unique properties. The present Special Issue is then focused on such last topics, collecting eight research papers and one review article dealing with MoS_2 [1], graphene [2–4] and other 2D nanomaterials integrated onto several kinds of materials and structures (nanogap [1], porous silicon [5], silicon carbide [6]) and for different applications (photoluminescence [1], ink-jet printing [7], optics and plasmonics [2,8], lubrication [3], innovative patterning [4], biomedicine [9]).

In particular, Yang et al. [1] proposed an innovative solution to bypass the limitation of the atomic thickness of monolayer MoS_2 hindering its optical absorption and emission in view of optoelectronic applications. By integrating monolayer MoS_2 onto nanometer wide gold nanogap arrays, it was possible to exploit the associated plasmon resonance, thus enhancing photoluminescence by a factor 20x, and thus paving the way for successive applications in photodetectors, sensors and emitters. Zhang J. et al. [2] studied another hybrid metamaterial consisting in metal–graphene. Such a coupling gives rise to a sharper Fano resonance than the pure graphene and it is, moreover, adjustable as a function of the number of graphene layers. Plasmonic sensing applications with extremely high sensitivity in the mid-infrared range are envisioned. Zhang L. et al. [3] showcase an interesting and relatively unconventional application for graphene, i.e., lubrication for structural ceramics. In particular, Si_3N_4/ Si_3N_4 sliding pair tribological properties (lubrication and cooling) have been investigated through the addition of different weight contents of graphene to a base lubricating oil. Relevant results have been obtained and an explanation about the lubricating improvement mechanism was provided. In the last paper involving graphene, Verna et al. [4] showed an original and straightforward process, based on the lift-off of the catalyst seed layer, to pattern few-layer graphene. The direct chemical vapor deposition of graphene on the patterned seed layer guaranteed high quality of the resulting 2D material and a 10 μm patterning resolution was demonstrated.

Volovlikova et al. [5] analyzed the effect of illumination intensity and p-dopant concentration on the dissolution properties of silicon for its photo-assisted etching with no external bias or metals to produce porous silicon. A thorough characterization was performed, providing valuable data for the control of porous silicon thickness and porosity. Ruffino et al. [6] provided a valuable basic study on the growth and coalescence characteristics of a nanoscale-thick bimetallic film of Au/Pd on a silicon carbide substrate. The kinetic of the growth process was studied from the initial 3D clustering to the final continuous rough thin film formation. Xiao et al. [7] performed a full comparison, with special attention to electrical resistivity and adhesion, between ink-jet-printed silver thin films based on nanoparticle ink and metal–organic decomposition ink cured by two different approaches, that is to say, UV exposure and heat-assisted approaches. Cao et al. [8] fabricated (through a combination of lithography and nanoimprint technology) and characterized a flexible 3D display film element consisting of two integrated structures of a microimage array and microlens array.

Finally, in their review article, Coluccio et al. [9] revised and critically described nanoscale transport phenomena and biomedical applications of different emerging electrochemical devices whose working principle relies on the interaction between ions and conductive polymers.

Last, but not least, we would like to thank all the contributing authors and all the involved reviewers for their precious contribution to the assembly and quality of this Special Issue.

Conflicts of Interest: The authors declare no conflict of interest.

References

1. Yang, Y.; Pan, R.; Tian, S.; Gu, C.; Li, J. Plasmonic Hybrids of MoS_2 and 10-nm Nanogap Arrays for Photoluminescence Enhancement. *Micromachines* **2020**, *11*, 1109. [CrossRef] [PubMed]
2. Zhang, J.; Hong, Q.; Zou, J.; He, Y.; Yuan, X.; Zhu, Z.; Qin, S. Fano-Resonance in Hybrid Metal-Graphene Metamaterial and Its Application as Mid-Infrared Plasmonic Sensor. *Micromachines* **2020**, *11*, 268. [CrossRef] [PubMed]
3. Zhang, L.; Wei, X.; Wang, J.; Wu, Y.; An, D.; Xi, D. Experimental Study on the Lubrication and Cooling Effect of Graphene in Base Oil for Si_3N_4/Si_3N_4 Sliding Pairs. *Micromachines* **2020**, *11*, 160. [CrossRef] [PubMed]
4. Verna, A.; Marasso, S.; Rivolo, P.; Parmeggiani, M.; Laurenti, M.; Cocuzza, M. Lift-Off Assisted Patterning of Few Layers Graphene. *Micromachines* **2019**, *10*, 426. [CrossRef] [PubMed]
5. Volovlikova, O.; Gavrilov, S.; Lazarenko, P. Influence of Illumination on Porous Silicon Formed by Photo-Assisted Etching of p-Type Si with a Different Doping Level. *Micromachines* **2020**, *11*, 199. [CrossRef] [PubMed]
6. Ruffino, F.; Censabella, M.; Piccitto, G.; Grimaldi, M. Morphology Evolution of Nanoscale-Thick Au/Pd Bimetallic Films on Silicon Carbide Substrate. *Micromachines* **2020**, *11*, 410. [CrossRef] [PubMed]
7. Xiao, P.; Zhou, Y.; Gan, L.; Pan, Z.; Chen, J.; Luo, D.; Yao, R.; Chen, J.; Liang, H.; Ning, H. Study of Inkjet-Printed Silver Films Based on Nanoparticles and Metal-Organic Decomposition Inks with Different Curing Methods. *Micromachines* **2020**, *11*, 677. [CrossRef] [PubMed]
8. Cao, A.; Xue, L.; Pang, Y.; Liu, L.; Pang, H.; Shi, L.; Deng, Q. Design and Fabrication of Flexible Naked-Eye 3D Display Film Element Based on Microstructure. *Micromachines* **2019**, *10*, 864. [CrossRef] [PubMed]
9. Coluccio, M.; Pullano, S.; Vismara, M.; Coppedè, N.; Perozziello, G.; Candeloro, P.; Gentile, F.; Malara, N. Emerging Designs of Electronic Devices in Biomedicine. *Micromachines* **2020**, *11*, 123. [CrossRef] [PubMed]

Article

Plasmonic Hybrids of MoS_2 and 10-nm Nanogap Arrays for Photoluminescence Enhancement

Yang Yang [1,†], Ruhao Pan [1,†], Shibing Tian [1], Changzhi Gu [1,2] and Junjie Li [1,2,3,*]

1 Beijing National Laboratory for Condensed Matter Physics, Institute of Physics, Chinese Academy of Sciences, P.O. Box 603, Beijing 100190, China; yang.yang@iphy.ac.cn (Y.Y.); panruhao@iphy.ac.cn (R.P.); tianshibing@iphy.ac.cn (S.T.); czgu@iphy.ac.cn (C.G.)
2 School of Physical Sciences, University of Chinese Academy of Sciences, Beijing 100049, China
3 Songshan Lake Materials Laboratory, Dongguan 523808, China
* Correspondence: jjli@iphy.ac.cn
† These authors contributed equally to this work.

Received: 17 October 2020; Accepted: 10 December 2020; Published: 15 December 2020

Abstract: Monolayer MoS_2 has attracted tremendous interest, in recent years, due to its novel physical properties and applications in optoelectronic and photonic devices. However, the nature of the atomic-thin thickness of monolayer MoS_2 limits its optical absorption and emission, thereby hindering its optoelectronic applications. Hybridizing MoS_2 by plasmonic nanostructures is a critical route to enhance its photoluminescence. In this work, the hybrid nanostructure has been proposed by transferring the monolayer MoS_2 onto the surface of 10-nm-wide gold nanogap arrays fabricated using the shadow deposition method. By taking advantage of the localized surface plasmon resonance arising in the nanogaps, a photoluminescence enhancement of ~20-fold was achieved through adjusting the length of nanogaps. Our results demonstrate the feasibility of a giant photoluminescence enhancement for this hybrid of MoS_2/10-nm nanogap arrays, promising its further applications in photodetectors, sensors, and emitters.

Keywords: monolayer MoS_2; 10-nm nanogap; localized surface plasmon resonance; photoluminescence

1. Introduction

In the past ten years, two-dimensional (2D) transition metal dichalcogenides (TMDs) have received plenty of research interest, due to their striking physical properties and applications in optoelectronic devices [1,2]. Molybdenum disulphide (MoS_2) is a representative member of the TMDs family [3,4], in which the bandgap can transit from indirect to direct [5,6], when the thickness is reduced to a monolayer. The bandgap shifts from 1.29 eV for the bulk MoS_2 to 1.9 eV for the monolayer MoS_2, accompanied with an enhancement of the photoluminescence (PL) up to 10^4 [5]. Therefore, the direct-bandgap characteristic of the monolayer MoS_2 leads to attractive applications in phototransistors [7], photodetectors [8], light emitters [9], and photocatalysis [10]. However, the thickness of 2D MoS_2 is too thin to absorb sufficient light, which limits the light-harvest efficiency and consequently restricts its practical applications. Therefore, efficiently enhancing the light absorption and photoluminescence (PL) emission of MoS_2 has become an important issue for exploring the practical applications in optoelectronic devices. In recent years, integrating MoS_2 with plasmonic nanoscale metals has been demonstrated to be an effective route to promote the optical properties of MoS_2 [11–13].

Plasmonic nanoscale metals, including noble metal nanoparticles and nanostructures, can strongly enhance the electromagnetic (EM) fields of the excitation light, due to the localized surface plasmon resonance (LSPR) on the surface of nanoscale metals [14–16]. Gao et al. have prepared the hybrids of MoS_2 and Ag nanoparticles, including shape-controlled cubes, octahedra, and spherical particles, and systematically studied the influences of the morphology of nanoparticles on the PL emission of MoS_2 [17]. Compared with nanoparticles, nanostructures fabricated using advanced nanofabrication techniques exhibit higher controllability, reproducibility, and large-scale periodicity [18]. By modulating the diameter of nanodiscs, the 12-times PL enhancement was achieved in the hybrids of nanodisc arrays and monolayer MoS_2 [19]. Among the various species of nanostructures, nanogaps have exhibited a prominent PL enhancement due to the EM field and the strong LSPR effect in the nanogap zone. The EM field intensity is strengthened as the nanogap size decreases, especially when the width of nanogaps is smaller than 10 nm [20,21]. Wang et al. reported a giant PL enhancement up to 20,000-fold for the hybrid of WSe_2 on 12-nm nanotrenches [22]. In 2018, Cai et al. proposed a hybrid of $MoSe_2$ on large area ultranarrow annular nanogap arrays (ANAs), which were fabricated using atomic layer deposition and polystyrene spheres lithography techniques [23]. Nanofabrication techniques, such as E-beam lithography (EBL) and focused ion beam (FIB) milling have been explored to directly fabricate 10-nm nanogap arrays [24–26]. However, the fabrication processes based on these techniques are relatively complex, due to fact that the small size of the nanogap is close to the limitation of resolution. In 2018, Hao et al. proposed a hybrid of MoS_2 and patterned plasmonic dimers fabricated by a facile approach utilizing porous anodic aluminium oxide (AAO) templates during the angle-resolved shadow deposition [27]. The shadow deposition method, which is based on the inclined deposition of materials on the prefabricated pattern, has been demonstrated as a feasible way to fabricate 10-nm nanogap arrays over a large area [28,29]. However, the MoS_2 hybrids based on 10-nm nanogap arrays fabricated by the shadow deposition method is still rarely reported.

In this work, a type of plasmonic hybrid composed of 10-nm Au nanogap arrays and monolayer MoS_2 was proposed for PL enhancement. The 10-nm Au nanogap arrays were fabricated using the shadow deposition method, which was composed by depositing nanostrips with a 20 degree inclining angle on the nanostrips previously fabricated. By adjusting the length of nanogaps, the PL enhancement can be significantly boosted up to ~20-fold for the MoS_2/nanogaps hybrid formed with 240-nm-length nanogaps under the excitation of a 532-nm laser. Combined with the finite-different time-domain (FDTD) simulation, the mechanism behind the PL enhancement was analyzed. Our results provide a feasible method to prepare large area MoS_2-nanostructure plasmonic hybrids with a giant photoluminescence enhancement, promising their further applications in photodetectors, sensors, and emitters.

2. Materials and Methods

The monolayer MoS_2 used in this work was fabricated on a single-crystalline sapphire substrate using the chemical vapor deposition (CVD) method [30]. The 10-nm nanogap arrays were fabricated using the shadow deposition method, a combination of EBL and electron beam deposition (EBD) techniques. Figure 1 illustrates the fabrication processes for the hybrid of MoS_2 and nanogap arrays, which mainly include six steps. First, as presented in Figure 1a, a polymethyl methacrylate (PMMA) 495 A5 layer with a 200-nm thickness was spin coated onto the Si substrate, and the film was pre-baked on a hot plate at 180 °C for 60 s. Then, the EBL was used to write the patterns of nanostrips along the y-axis. Third, metallic nanostrips can be transferred by EBD on a 60-nm Au film and followed lift-off processes. After that, the nanostrips with a 60-nm-height were observed, as presented in Figure 1d. Fourth, the nanostrips along the x-axis were written by the alignment lithography, as illustrated in Figure 1e. Then, the shadow deposition of Au with a 20 degree inclining angle was applied to the Au strips formed in step 3, the thickness of Au was also 60-nm in the shadow deposition process, as shown in Figure 1f. Since the shadow origins from the deposition angle and steps, 10-nm nanogaps can be obtained, as illustrated in Figure 1g. Finally, the MoS_2 was transferred on top of the nanogap arrays, forming a plasmonic MoS_2/nanogaps hybrid, as shown in Figure 1h. The MoS_2 flake was removed from the substrate to Au nanogap arrays using a PMMA-assisted wet transfer procedure, which has been applied to transfer 2D materials [22,31]. In order to investigate the plasmonic enhancement effect of the Au nanogaps on the PL emission of MoS_2, another type of MoS_2-nanogaps hybrid, in which the nanogap arrays were fabricated directly on the CVD-grown MoS_2 on the sapphire substrate (nanogaps/MoS_2) was fabricated for comparison.

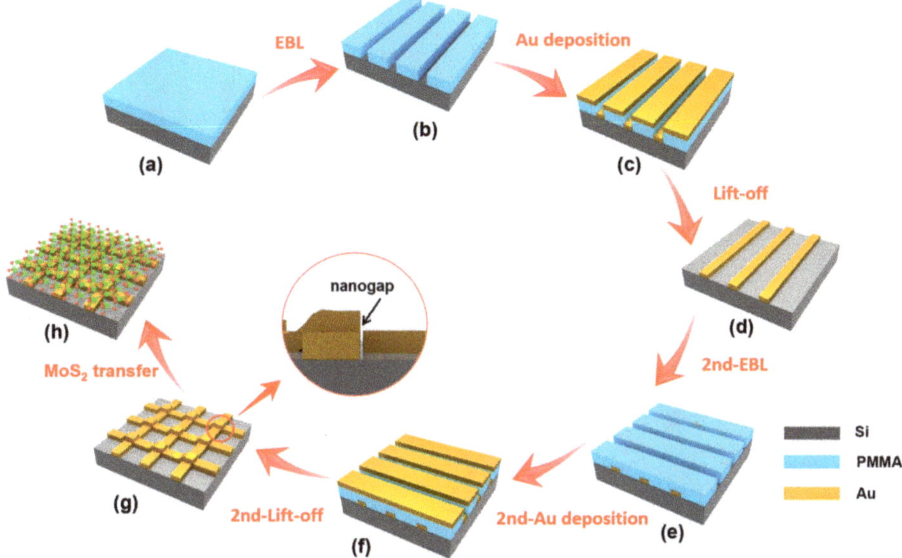

Figure 1. Schematic fabrication procedures of the MoS_2/Au nanogap hybrid. (**a**) Substrate coated with polymethyl methacrylate (PMMA). (**b**) Nanotranches formed after the first E-beam lithography (EBL) process. (**c**) First Au deposition process. (**d**) Au nanostrips obtained after lift-off process. (**e**) Nanotrenches formed after the second EBL process. (**f**) The second Au shadow deposition process. (**g**) Nanogaps obtained by crossover of nanostrips. (**h**) Hybrid structure formed after MoS_2 transferred. An enlarged view of the nanogap obtained was illustrated in the middle.

The morphology and dimensions of the fabricated Au nanogaps are detected using a scanning electron microscope (SEM). Figure 2a presents the SEM images of the fabricated 10-nm Au nanogap arrays with a periodicity of 1 μm. It can clearly be seen that the nanogaps formed on the right side on each cubic island. Since the nanogaps were formed as a result of the shadow of first-layer nanostrips, the width of the nanogap obviously depended on the inclining angle for the deposition of the second-layer nanostrips. As exhibited in Figure 2b, the gap width widens with increasing the inclined deposition angle. The 10-nm nanogaps have been successfully fabricated by fixing the inclining angle at 20 degrees, as shown in Figure 2c. The length of this nanogap was determined as ~240 nm, since the line-width of the nanostrips was set at 240 nm. Therefore, the length of the nanogaps can be feasibly modulated by adjusting their line-width. Figure 2d shows the SEM image of the monolayer MoS_2 transferred onto the nanogap arrays. It can be seen that the nanogap arrays are fully covered with the MoS_2 monolayer.

Photoluminescence (PL) and Raman measurements were carried on a micro-confocal Raman spectrometer (Horiba HR Evolution) equipped with a microscope (BX41, Olympus, Tokyo, Japan). A 100 × (NA = 0.9) objective lens was used for focusing the laser on the sample surface and collecting the PL signal. The 532-nm laser was used as the excitation sources for PL measurements and the laser power on the sample was about 0.5 mW in order to prevent the overheating effect. The spectra acquisition time was 30 and 10 s for Raman and PL acquisition, respectively. Figure 3a exhibits the schematic setup of the PL measurements on the MoS_2/nanogap hybrids. As illustrated in Figure 3a, the laser beam was a normal incident on the sample surface, and the polarization direction of the incident laser was along the *x*-axis. In other words, the PL signal was collected while the incident polarization was perpendicular to the direction of the nanogap.

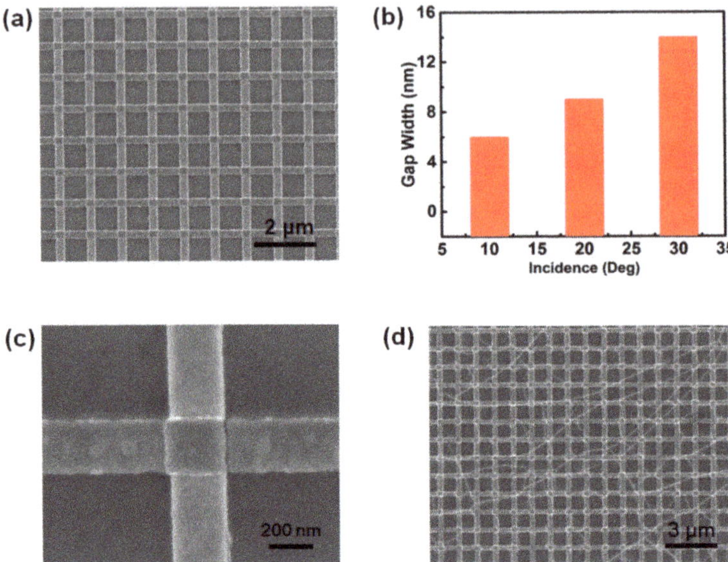

Figure 2. (**a**) scanning electron microscope (SEM) image of 10-nm nanogap arrays fabricated using the shadow deposition method. (**b**) Gap width obtained as a function of the inclining angle. (**c**) Magnified SEM image of a 10-nm nanogap of 240-nm-length. (**d**) SEM image for the nanogap arrays covered with MoS_2.

In order to solve the field enhancement effect of the nanogaps, the finite-different time-domain (FDTD) method, which was a state-of-the-art method for solving Maxwell's equations in complex geometries, had been widely used in the nanogap configurations to collect the near-field distribution, transmission/reflection [23,32]. Here, in this work, the FDTD method was employed to simulate the electric field distribution on the 10-nm nanogap arrays. The geometries of the Au nanogaps in the simulations were designed to match the SEM images shown in Figure 2. The nanogaps chosen for the simulations were combined with steps with a 60-nm-height and the gaps that were built up by the shadow deposition with the width of the gaps were set to 10 nm. In addition, the gap length was determined by the linewidth of the nanostrip, which was used in the shadow deposition, and nanogaps with different lengths were simulated. The wavelength of the incidence light was set at 532 nm and propagated along the z-axis with the electric field polarized along the x- and y-axes, respectively. The source was a 1×1 μm Gaussian wave, which was similar to the excited laser used in the experiment. All the simulated boundaries were 12 layers perfectly matched layers (PML) to avoid reflections, and all the mesh steps along the x, y, and z-axes are set to 2 nm in order to obtain accurate results of the filed distribution in the nanogaps. Two monitors were placed perpendicular to capture the field distribution, one of them was perpendicular to the z-axes and overlapped the top surface of the structure, while the other was perpendicular to the y-axes and was placed in the nanogap. The filed distributions of the gaps with different lengths were collected.

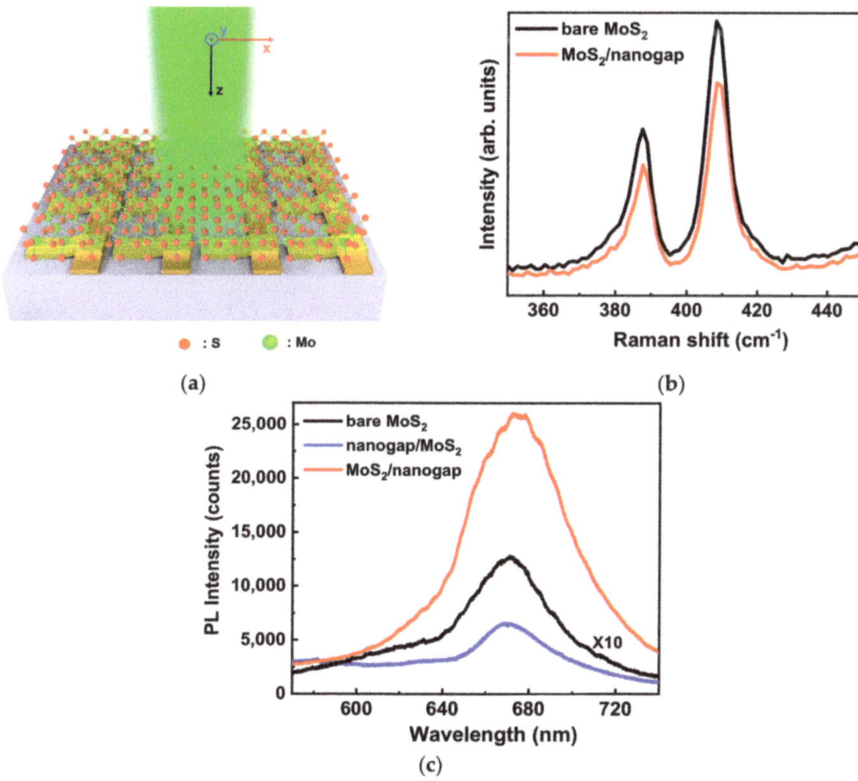

Figure 3. (**a**) Schematic illustration of the photoluminescence (PL) measurements on MoS_2/nanogap hybrids. (**b**) Raman spectra for the bare MoS_2 and MoS_2/nanogap hybrid, respectively. (**c**) Typical PL spectra for the MoS_2/nanogap hybrid, nanogap/MoS_2 hybrid, and bare MoS_2, respectively, collected under a 532-nm excitation.

3. Results and Discussion

Figure 3b presents the Raman spectra collected from a bare MoS_2 and a hybrid MoS_2/nanogap, respectively, which resembles the typical features of MoS_2. It is worth noting that there is no obvious difference, such as frequency shift and peak broadening, between these two spectra suggesting that the hybridization with Au nanogaps did not introduce additional local strains or defects in the top MoS_2 layer [33]. Figure 3c presents the typical PL spectra from two types of MoS_2-nanogap hybrids (MoS_2/nanogap and nanogap/MoS_2) excited under a 532 nm laser. The length of the selected nanogap in the hybrids was 240 nm. The line shape of the PL spectra for the MoS_2-nanogap hybrids are almost the same as that for the bare MoS_2 (black curve), composed with an intense peak around 670 nm and a weak shoulder peak around 620 nm, which is consistent with those reported in the literature [34–36]. These two PL peaks are attributed as the A and B excitons, respectively, corresponding to the direct gap transition at the K point. Their energy difference is due to the spin-orbital splitting of the valence band. The A exciton for the bare MoS_2 was centered at 670 nm, whereas the A exciton for the MoS_2/nanogap hybrid red-shifts at 675 nm. One can see that the PL intensity from these two plasmonic hybrids displayed an obvious enhancement compared with the bare MoS_2. Remarkably, the MoS_2/nanogap hybrid exhibited an extraordinary PL enhancement with an amplitude up to 20-fold compared with the bare MoS_2. In contrast, the PL intensity was just enhanced by five times while fabricating nanogaps arrays on top of MoS_2. These results demonstrate that the MoS_2/nanogap hybrid possesses a much better PL enhancement effect than the nanogap/MoS_2 hybrid.

Figure 4a presents the enhanced PL spectra for the MoS_2/nanogap hybrids with lengths of 200, 220, 240, and 260 nm, respectively. It is noteworthy that the PL intensity of the MoS_2/nanogap hybrids boosted with the increased line width of the nanostrips, and achieved the highest value for the 240-nm nanogap length. Then, the intensity of PL decreased quickly while the nanogap length increased to 260 nm. Using the PL intensity of the bare MoS_2 as the reference, the PL enhancement of the hybrids was calculated as a function of the nanogap length. As shown in Figure 4b, the PL emission from the hybrids was increased from 7 to ~20 times, while the nanogap length increased from 200 to 240 nm, compared with the bare MoS_2. With the increased nanogap length, the PL enhancement falls to 5-folds. The results in Figure 4 clearly demonstrate that the PL intensity enhancement is closely associated with the length of the nanogaps.

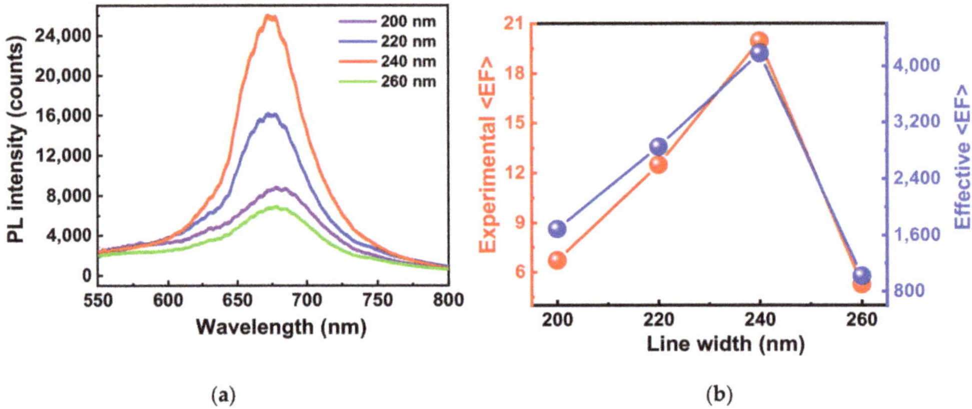

Figure 4. (**a**) Enhanced PL spectra for the MoS_2/nanogap hybrids constituted of nanogaps with different lengths. (**b**) Experimental and effective PL enhancement factors as a function of the nanogap length.

To see the intensity enhancement distribution on the plasmonic hybrid nanostructure, the PL mapping measurement was performed on a $10 \times 10\ \mu m^2$ area selected from the hybrid composed of the monolayer MoS_2 and 240-nm-length nanogap arrays, as shown in Figure 5a. The mapping image was created by integrating the PL intensities over the range of 630–720 nm. As presented in Figure 5b, a gridding-like image for the PL distribution is observed, implying that the highest PL intensities appear at the intersections. The periodicity for meshes is 1 μm, which is consistent with that for the nanogaps fabricated in this work. This demonstrates that the highest PL intensities take place on the crossover of first- and second-layer nanostrips. This is attributed to the hot spots formed on the vertices of nanogaps due to the LSPR effect excited under a 532-nm laser. However, one cannot identify where the hot spots are specifically located in the nanogap, as the diameter of the laser spot is approximately 800 nm, which is much larger than the width (10 nm) and length (240 nm) of the nanogap. For future investigating the distribution of the hot spot in nanogaps, the probing technique with a higher spatial resolution, such as tip-enhanced Raman spectroscopy (TERS), is needed. The results in Figure 5 demonstrate that the nanogap arrays fabricated in this work were able to enhance the PL emission of MoS_2 prominently. Moreover, the nanogap arrays were fully covered by the monolayer MoS_2. The differences in the enhanced PL intensity between the nanogaps could be attributed to the defects or strains generated during the MoS_2 transferring procedure.

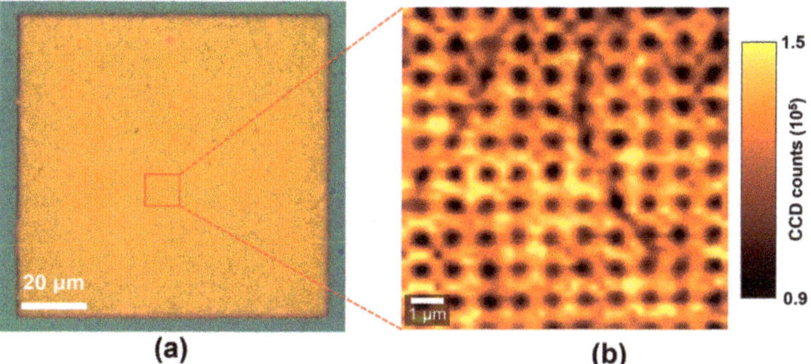

Figure 5. (**a**) Optical microscope image and (**b**) PL mapping image for the hybrid composed of the monolayer MoS_2 and 240-nm-length nanogap arrays.

In order to understand the physical mechanism behind the PL enhancement for the MoS_2/nanogap hybrid, the FDTD simulation was employed to evaluate the electromagnetic field distribution of the 10-nm Au nanogap under a 532-nm excitation. Figure 6 shows the map of the simulated electric field enhancement $|E/E_0|$ on the surface of the nanogap formed of 240-nm-width nanostrips under x-polarization. E is the local electric field at the surface and E_0 is the incident electric field. One can see in Figure 6a that the maximum enhancement of the electric field takes place at the two vertices of the second layer nanostrip, which is ~8 times under the 532-nm excitation. In contrast, the electric field at the interface between the first layer nanostrip and substrate does not exhibit an obvious enhancement, as exhibited in Figure 6b. Figure 6c shows the field distribution in the x-z plane, in which one can see that the hot spots do not exhibit the same amplitude on the two sites of the nanogap. A bright hot spot emerges at the vertices of the second layer on the left side, while a weaker spot appears at the vertices of the second layer on the right side. This may be attributed to the height difference of 60 nm between the two sides of the nanogap, so that the plasmon coupling in the nanogap zone is not prominent. As demonstrated in previous literatures, the literal dimension of the nanogap is normally very small, for example, nanoparticle dimer or nanotips, so that the EM field can be confined in a limited space and significantly enhanced. The interaction of plasmons along the interparticle axis of a nanoparticle dimer results in low-energy longitudinal bonding dipole plasmon (LBDP) modes [37,38]. However, the length of the nanogap in our work is too long, the surface plasmon can propagate along the edges of the nanogap. The EM field cannot uniformly be distributed on the whole nanogap zone, and the highest EM field (hot spots) only takes place on the vertices of the nanogaps. Therefore, the hot spots on the 240-nm-length nanogap primarily arise from the LSPR rather than the plasmon coupling modes.

The results in Figure 6 clearly explain why the PL intensity for the MoS_2/nanogap hybrid is much higher than that for the nanogap/MoS_2 hybrid (see Figure 3c). The discrepancy in the PL enhancement between these two types of hybrid structures could be attributed to their own architectures, in other words, the location of the monolayer MoS_2 in the hybrids. As it has been demonstrated, the EM field enhancement is a type of near field effect, which requires the molecule to be within a few nanometers from the surface of the plasmonic nanostructure for an appreciable enhancement in order to be obtained [39,40]. For the MoS_2/nanogap hybrid structure, the monolayer MoS_2 is transferred on top of the nanogap. The hot spots excited by the 532-nm laser can directly interact with MoS_2, consequently enhancing the PL emission of MoS_2. In sharp contrast, MoS_2 is under the nanogap structures in the nanogap/MoS_2 hybrid structure, which is 60-nm far away from the hot spots, the LSPR cannot interact with the MoS_2 layer, so the PL intensity of MoS_2 cannot be enhanced obviously.

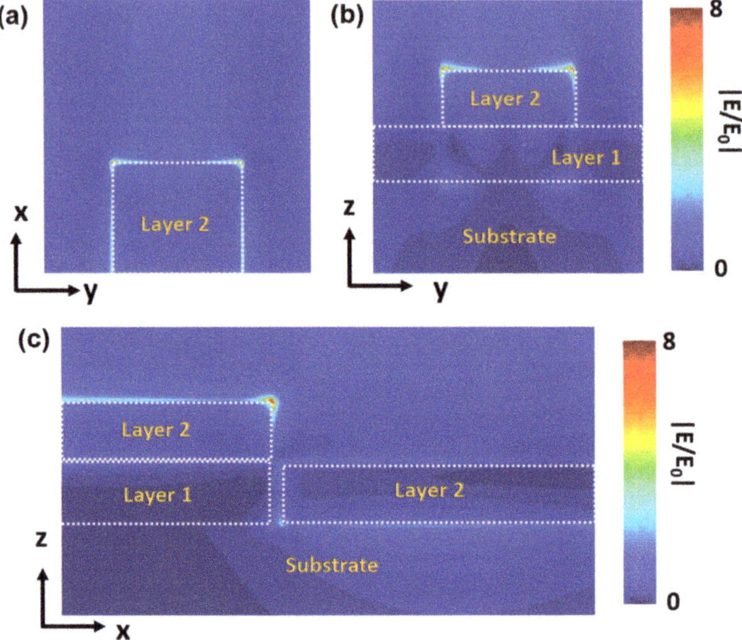

Figure 6. (**a**) Top view and cross-sectional view for the (**b**) z-y plane and (**c**) x-z plane of the calculated electric field distribution on the 10-nm Au nanogap with 240-nm-length excited by a 532-nm laser.

Figure 7 presents the cross-sectional view of the simulated electric field enhancement $|E/E_0|$on the nanogaps with a length of 200, 220, 240, 260, and 280 nm, respectively. One can see that the maximum electric field enhancement ("hot spots") appears at the top two corners of the second layer for all the nanogaps. Noteworthily, the maximum electromagnetic field enhancement takes place at the nanogap with a 240-nm-length. However, the difference of the field enhancement between these nanogaps is not obvious. Therefore, the enhancement of the light at the hot spots was calculated using $|E/E_0|^2$ and presented in Figure 7f. As illustrated clearly in Figure 7f, the light (near field) was enhanced by ~64-fold for the nanogap with a 240-nm-length, whereas the field EF is down to 40-fold for the 280-nm nanogap length.

The simulation results in Figure 7f imply the significant light enhancement ability of these nanogap nanostructures. However, the maximum measured PL enhancement for the MoS_2/nanogap hybrids is just 20 times (see Figure 4b), which is lower than the simulated results. This phenomenon could be attributed to the architecture of the hybrids, in which only the nanogap area has contributions to the PL enhancement of MoS_2, as exhibited in Figure 7. In order to evaluate the PL enhancement of these nanogaps, the PL enhancement factor was corrected using the area of nanogaps. We calculated the effective average PL enhancement factor, ⟨EF⟩, using the following formula [22,41,42]:

$$\langle \mathrm{EF} \rangle = \frac{I_{hybrid}}{I_{bare}} \frac{A_{bare}}{A_{hybrid}} \tag{1}$$

where I_{hybrid} is the PL intensity from the MoS_2/nanogap hybrids and I_{bare} is the PL intensity from the bare MoS_2. A_{bare} represents the excitation area of the laser spot size ($\pi \times 400^2$ nm^2) and A_{hybrid} represents the area of the nanogaps ($10 \times$ gap length nm^2) within the laser spot, which depends on the length of nanogaps. The effective ⟨EF⟩ were calculated using Equation (1) and plotted as a function of the length of nanogaps in Figure 4b. Remarkably, the estimated ⟨EF⟩ reaches a factor up to ~4180 for the hybrid with 240-nm-length nanogap. The maximum effective EF is much higher than that of the simulated EM EF (~64). This is due to the fact that the effective PL enhancement is actually composed of both the plasmon-enhanced excitation process and the plasmon-enhanced emission process [22]. The simulation results in Figure 7f just present the EF for the excitation process. If the EF for the plasmon-enhanced emission is taken into consideration, the simulated EF would be at the same level of the effective EF.

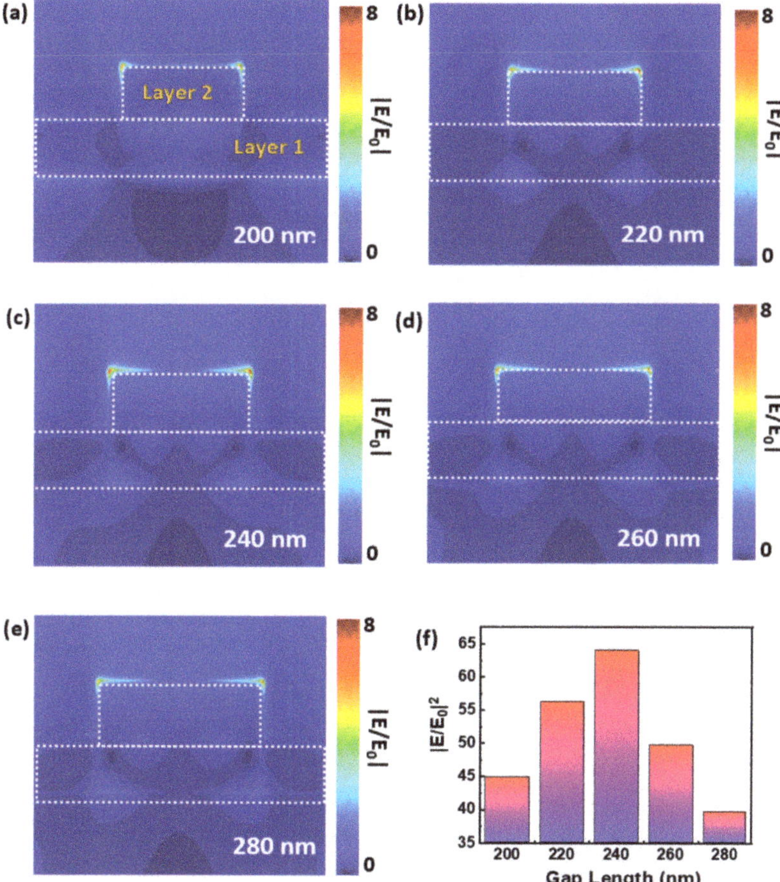

Figure 7. (**a**–**e**) Cross-sectional view of the simulated electric field distribution on the 10-nm Au nanogaps with selected gap lengths. (**f**) Calculated electric field enhancement of the 10-nm Au nanogap as a function of gap length.

Moreover, the effective ⟨EF⟩ decreased with the increasing gap length, and down to 210 for the 280-nm-length nanogap, which follows the similar trace as the experimental ⟨EF⟩, as shown in Figure 4b. The huge discrepancy between the experimental and effective ⟨EF⟩ should be attributed to the small effective interaction area between the nanogap and MoS_2. As shown in Figure 6, only the top corners of the nanogap exhibit a high electromagnetic field enhancement, whereas the other part of the nanostrips does not make dominating contributions to the PL enhancement. In order to easily compare the PL enhancement for the hybrids, the periodicity of the nanogap arrays is the same (1 μm) for all the samples used in this work. In other words, there is only 1 nanogap within the laser spot. Therefore, in the future device fabrication process, increasing the density of nanogaps would be a proper route to enhance the PL emission of MoS_2.

4. Conclusions

In summary, we have proposed a type of plasmonic hybrid for enhancing the PL emission of MoS_2 by fabricating 10-nm-wide Au nanogap arrays on the monolayer MoS_2. By taking advantage of the LSPR arising in the nanogaps, the PL emission of MoS_2 was significantly enhanced under a 532-nm excitation, which can be modulated by adjusting the width of the nanostrips. The maximum PL enhancement was demonstrated on the MoS_2/nanogaps hybrid formed with nanogaps of 240-nm-length, in which an effective emission enhancement of ~20-fold is attained. The effective enhancement factor for the MoS_2/nanogaps hybrid has achieved ~4180. The mechanism for the PL enhancement of the MoS_2/nanogaps hybrids was studied using the FDTD simulation. Our results demonstrate the feasibility of a giant photoluminescence enhancement for this hybrid of MoS_2/10-nm nanogaps, promising their further applications in photodetectors, sensors, and emitters.

Author Contributions: Design, fabrication, methodology, R.P.; writing—review and editing, Y.Y.; resources, S.T.; supervision, C.G. and J.L.; project administration, J.L.; funding acquisition, C.G. and J.L. All authors have read and agreed to the published version of the manuscript.

Funding: This research was funded by the National Key R&D Program of China (grant no. 2016YFA0200800, 2018YFB0703500 and 2016YFA0200400), the National Natural Science Foundation of China (grant no. 11674387, 12074420, 11574369, 61905274, 91323304, and 11704401), and the Key Research Program of Frontier Sciences of CAS (grant no. QYZDJ-SSW-SLH042).

Conflicts of Interest: The authors declare no conflict of interest.

References

1. Radisavljevic, B.; Radenovic, A.; Brivio, J.; Giacometti, V.; Kis, A. Single-layer MoS_2 transistors. *Nat. Nanotechnol.* **2011**, *6*, 147–150. [CrossRef] [PubMed]
2. Zheng, W.; Jiang, Y.; Hu, X.; Li, H.; Zeng, Z.; Wang, X.; Pan, A. Light emission properties of 2D transition metal dichalcogenides: Fundamentals and applications. *Adv. Opt. Mater.* **2018**, *6*, 1800420. [CrossRef]
3. Wang, Q.H.; Kalantar-Zadeh, K.; Kis, A.; Coleman, J.N.; Strano, M.S. Electronics and optoelectronics of two-dimensional transition metal dichalcogenides. *Nat. Nanotechnol.* **2012**, *7*, 699–712. [CrossRef] [PubMed]
4. Low, T.; Chaves, A.; Caldwell, J.D.; Kumar, A.; Fang, N.X.; Avouris, P.; Heinz, T.F.; Guinea, F.; Martin-Moreno, L.; Koppens, F. Polaritons in layered two-dimensional materials. *Nat. Mater.* **2017**, *16*, 182–194. [CrossRef]
5. Mak, K.F.; Lee, C.; Hone, J.; Shan, J.; Heinz, T.F. Atomically Thin MoS_2: A New Direct-Gap Semiconductor. *Phys. Rev. Lett.* **2010**, *105*, 136805. [CrossRef]
6. Liu, Y.; Nan, H.; Wu, X.; Pan, W.; Wang, W.; Bai, J.; Zhao, W.; Sun, L.; Wang, X.; Ni, Z. Layer-by-Layer Thinning of MoS_2 by Plasma. *ACS Nano* **2013**, *7*, 4202–4209. [CrossRef]
7. Yin, Z.; Li, H.; Li, H.; Jiang, L.; Shi, Y.; Sun, Y.; Lu, G.; Zhang, Q.; Chen, X.; Zhang, H. Single-layer MoS_2 phototransistors. *ACS Nano* **2012**, *6*, 74–80. [CrossRef]
8. Lopez-Sanchez, O.; Lembke, D.; Kayci, M.; Radenovic, A.; Kis, A. Ultrasensitive photodetectors based on monolayer MoS_2. *Nat. Nanotechnol.* **2013**, *8*, 497–501. [CrossRef]
9. Nan, H.; Wang, Z.; Wang, W.; Liang, Z.; Lu, Y.; Chen, Q.; He, D.; Tan, P.; Miao, F.; Wang, X.; et al. Strong photoluminescence enhancement of MoS_2 through defect engineering and oxygen bonding. *ACS Nano* **2014**, *8*, 5738–5745. [CrossRef]
10. Wang, J.; Yan, M.; Zhao, K.; Liao, X.; Wang, P.; Pan, X.; Yang, W.; Mai, L. Field Effect Enhanced Hydrogen Evolution Reaction of MoS_2 Nanosheets. *Adv. Mater.* **2017**, *29*, 1604464. [CrossRef]
11. Zeng, Y.; Li, X.; Chen, W.; Liao, J.; Lou, J.; Chen, Q. Highly Enhanced Photoluminescence of Monolayer MoS_2 with Self-Assembled Au Nanoparticle Arrays. *Adv. Mater. Interfaces* **2017**, *4*, 1700739. [CrossRef]
12. Zu, S.; Li, B.; Gong, Y.; Li, Z.; Ajayan, P.M.; Fang, Z. Active Control of Plasmon–Exciton Coupling in MoS_2–Ag Hybrid Nanostructures. *Adv. Opt. Mater.* **2016**, *4*, 1463–1469. [CrossRef]
13. Shi, W.; Zhang, L.; Wang, D.; Zhang, R.; Zhu, Y.; Zhang, L.; Peng, R.; Bao, W.; Fan, R.; Wang, M. Hybrid coupling enhances photoluminescence of monolayer MoS_2 on plasmonic nanostructures. *Opt. Lett.* **2018**, *43*, 4128–4131. [CrossRef] [PubMed]

14. Ding, S.Y.; You, E.M.; Tian, Z.Q.; Martin, M. Electromagnetic theories of surface-enhanced Raman spectroscopy. *Chem. Soc. Rev.* **2017**, *46*, 4042–4076. [CrossRef]
15. Xu, Y.; Ji, D.X.; Song, H.M.; Zhang, N.; Hu, Y.W.; Thomas, D.A.; Enzo, M.D.; Fabrizio, S.X.; Gan, Q.Q. Light–Matter Interaction within Extreme Dimensions: From Nanomanufacturing to Applications. *Adv. Opt. Mater.* **2018**, 1800444. [CrossRef]
16. Denis, G.B.; Martin, W.; Jorge, C.; Tomasz, J.A.; Timur, S. Novel Nanostructures and Materials for Strong Light–Matter Interactions. *ACS Photonics* **2018**, *5*, 24–42.
17. Gao, W.; Lee, Y.H.; Jiang, R.; Wang, J.; Liu, T.; Ling, X.Y. Localized and continuous tuning of monolayer MoS_2 photoluminescence using a single shape-controlled Ag nanoantenna. *Adv. Mater.* **2016**, *28*, 701–706. [CrossRef]
18. Joel, H.; Lee, J.; Lee, M.H.; Hasan, W.; Odom, T.W. Nanofabrication of Plasmonic Structures. *Annu. Rev. Phys. Chem.* **2009**, *60*, 147–165.
19. Serkan, B.; Sefaattin, T.; Koray, A. Enhanced Light Emission from Large-Area Monolayer MoS_2 Using Plasmonic Nanodisc Arrays. *Nano Lett.* **2015**, *15*, 2700–2704.
20. Nam, J.M.; Oh, J.W.; Lee, H.; Suh, Y.D. Plasmonic Nanogap-Enhanced Raman Scattering with Nanoparticles. *Acc. Chem. Res.* **2016**, *49*, 2746–2755. [CrossRef]
21. Yang, Y.; Gu, C.Z.; Li, J.J. Sub-5 nm Metal Nanogaps: Physical Properties, Fabrication Methods, and Device Applications. *Small* **2019**, *15*, 1804177. [CrossRef]
22. Wang, Z.; Dong, Z.; Gu, Y.; Chang, Y.H.; Zhang, L.; Li, L.J.; Zhao, W.; Eda, G.; Zhang, W.; Grinblat, G. Giant photoluminescence enhancement in tungsten-diselenide–gold plasmonic hybrid structures. *Nat. Commun.* **2016**, *7*, 11283. [CrossRef]
23. Cai, H.B.; Meng, Q.S.; Zhao, H.; Li, M.L.; Dai, Y.M.; Lin, Y.; Ding, H.Y.; Pan, N.; Tian, Y.C.; Luo, Y.; et al. High-Throughput Fabrication of Ultradense Annular Nanogap Arrays for Plasmon-Enhanced Spectroscopy. *ACS Appl. Mater. Interfaces* **2018**, *10*, 20189–20195. [CrossRef] [PubMed]
24. Duan, H.; Hu, H.; Kumar, K.; Shen, Z.; Yang, J.K.W. Direct and reliable patterning of plasmonic nanostructures with sub-10-nm gaps. *ACS Nano* **2011**, *5*, 7593–7600. [CrossRef] [PubMed]
25. Cui, A.; Liu, Z.; Dong, H.; Wang, Y.; Zhen, Y.; Li, W.; Li, J.; Gu, C.; Hu, W. Single grain boundary break junction for suspended nanogap electrodes with gapwidth down to 1–2 nm by focused ion beam milling. *Adv. Mater.* **2015**, *27*, 3002–3006. [CrossRef] [PubMed]
26. Pan, R.; Yang, Y.; Wang, Y.; Li, S.; Liu, Z.; Su, Y.; Quan, B.; Li, Y.; Gu, C.; Li, J. Nanocracking and metallization doubly defined large-scale 3D plasmonic sub-10 nm-gap arrays as extremely sensitive SERS substrates. *Nanoscale* **2018**, *10*, 3171–3180. [CrossRef]
27. Hao, Q.; Pang, J.; Zhang, Y.; Wang, J.; Ma, L.; Schmidt, O.G. Boosting the Photoluminescence of Monolayer MoS_2 on High-Density Nanodimer Arrays with Sub-10 nm Gap. *Adv. Opt. Mater.* **2018**, *6*, 1700984. [CrossRef]
28. Siegfried, T.; Ekinci, Y.; Solak, H.H.; Martin, O.J.F.; Sigg, H. Fabrication of sub-10 nm gap arrays over large areas for plasmonic sensors. *Appl. Phys. Lett.* **2011**, *99*, 263302. [CrossRef]
29. Siegfried, T.; Ekinci, Y.; Martin, O.J.F.; Sigg, H. Gap plasmons and near-field enhancement in closely packed sub-10 nm gap resonators. *Nano Lett.* **2013**, *13*, 5449–5453. [CrossRef]
30. Yu, H.; Liao, M.; Zhao, W.; Liu, G.; Zhou, X.J.; Wei, Z.; Xu, X.; Liu, K.; Hu, Z.; Deng, K.; et al. Wafer-Scale Growth and Transfer of Highly-Oriented Monolayer MoS_2 Continuous Films. *ACS Nano* **2017**, *11*, 12001–12007. [CrossRef]
31. Cheng, F.; Johnson, A.D.; Tsai, Y.; Su, P.; Hu, S.; Ekerdt, J.G.; Shih, C. Enhanced Photoluminescence of Monolayer WS_2 on Ag Films and Nanowire–WS_2–Film Composites. *ACS Photonics* **2017**, *4*, 1421–1430. [CrossRef]
32. Chen, X.S.; Park, H.R.; Matthew, P.; Piao, X.; Nathan, C.L.; Hyungsoon, I.; Yun, J.K.; Jae, S.A.; Kwang, J.A.; Namkyoo, P.; et al. Atomic layer lithography of wafer-scale nanogap arrays for extreme confinement of electromagnetic waves. *Nat. Nanotechnol.* **2013**, *4*, 2361. [CrossRef]
33. Yan, A.M.; Sun, Y.H.; Ye, H.Y.; Liu, K.; Wang, R.M. Probing Evolution of Local Strain at MoS_2-Metal Boundaries by Surface-Enhanced Raman Scattering. *ACS Appl. Mater. Interfaces* **2018**, *10*, 40246–40254.
34. Andrea, S.; Sun, L.; Zhang, Y.; Li, T.; Kim, J.; Chim, C.-Y.; Galli, G.; Wang, F. Emerging Photoluminescence in Monolayer MoS_2. *Nano Lett.* **2010**, *10*, 1271–1275.

35. Zhang, C.; Wang, H.; Chan, W.; Manolatou, C.; Rana, F. Absorption of light by excitons and trions in monolayers of metal dichalcogenide MoS_2: Experiments and theory. *Phys. Rev. B* **2014**, *89*, 205436. [CrossRef]
36. Eda, G.; Yamaguchi, H.; Voiry, D.; Fujita, T.; Chen, M.; Chhowallam, M. Photoluminescence from Chemically Exfoliated MoS_2. *Nano Lett.* **2011**, *11*, 5111–5116. [CrossRef]
37. Jung, H.; Cha, H.; Lee, D.; Yoon, S. Bridging the Nanogap with Light: Continuous Tuning of Plasmon Coupling between Gold Nanoparticles. *ACS Nano* **2015**, *9*, 12292–12300. [CrossRef] [PubMed]
38. Kim, M.; Kwon, H.; Lee, S.; Yoon, S. Effect of Nanogap Morphology on Plasmon Coupling. *ACS Nano* **2019**, *13*, 12100–12108. [CrossRef]
39. Compagnini, G.; Galati, C.; Pignataro, S. Distance dependence of surface enhanced Raman scattering probed by alkanethiol self-assembled monolayers. *Phys. Chem. Chem. Phys.* **1999**, *1*, 2351–2353. [CrossRef]
40. Sicelo, S.M.; Ryan, A.H.; Nicolas, L.; Anne, I.H.; Michael, O.M.; George, C.S.; Peter, C.S.; Richard, P.V.D. High-Resolution Distance Dependence Study of Surface-Enhanced Raman Scattering Enabled by Atomic Layer Deposition. *Nano Lett.* **2016**, *16*, 4251–4259.
41. Akselrod, G.M.; Ming, T.; Argyropoulos, C.; Hoang, T.B.; Lin, Y.X.; Ling, X.; Smith, D.R.; Kong, J.; Mikkelsen, M.H. Leveraging Nanocavity Harmonics for Control of Optical Processes in 2D Semiconductors. *Nano Lett.* **2015**, *15*, 3578–3584. [CrossRef] [PubMed]
42. Kim, J.H.; Lee, H.S.; An, G.H.; Lee, J.; Oh, H.M.; Choi, J.; Lee, Y.H. Dielectric Nanowire Hybrids for Plasmon-Enhanced Light–Matter Interaction in 2D Semiconductors. *ACS Nano* **2020**, *14*, 11985–11994. [CrossRef] [PubMed]

Article

Experimental Study on the Lubrication and Cooling Effect of Graphene in Base Oil for Si_3N_4/Si_3N_4 Sliding Pairs

Lixiu Zhang [1,2], Xiaoyi Wei [1,2,*], Junhai Wang [1,2], Yuhou Wu [1,3], Dong An [1,2] and Dongyang Xi [4]

1 School of Mechanical Engineering, Shenyang Jianzhu University, Shenyang 110168, China; zhanglixiu@sjzu.edu.cn (L.Z.); jhwang@sjzu.edu.cn (J.W.); wuyh@sjzu.edu.cn (Y.W.); andong@sjzu.edu.cn (D.A.)

2 Test and Analysis Center, Shenyang Jianzhu University, Shenyang 110168, China

3 National-Local Joint Engineering Laboratory of NC Machining Equipment and Technology of High-Grade Stone, Shenyang 110168, China

4 School of Material Science and Engineering, Shenyang Jianzhu University, Shenyang 110168, China; xidy12@mails.jlu.edu.cn

* Correspondence: weixywei@163.com; Tel.: +86-024-2469-0088

Received: 18 December 2019; Accepted: 22 January 2020; Published: 3 February 2020

Abstract: Recently, the engineering structural ceramics as friction and wear components in manufacturing technology and devices have attracted much attention due to their high strength and corrosion resistance. In this study, the tribological properties of Si_3N_4/Si_3N_4 sliding pairs were investigated by adding few-layer graphene to base lubricating oil on the lubrication and cooling under different experimental conditions. Test results showed that lubrication and cooling performance was obviously improved with the addition of graphene at high rotational speeds and low loads. For oil containing 0.1 wt% graphene at a rotational speed of 3000 $r \cdot min^{-1}$ and 40 N loads, the average friction coefficient was reduced by 76.33%. The cooling effect on Si_3N_4/Si_3N_4 sliding pairs, however, was optimal at low rotational speeds and high loads. For oil containing 0.05 wt% graphene at a lower rotational speed of 500 $r \cdot min^{-1}$ and a higher load of 140 N, the temperature rise was reduced by 19.76%. In addition, the wear mark depth would decrease when adding appropriate graphene. The mechanism behind the reduction in friction and anti-wear properties was related to the formation of a lubricating protective film.

Keywords: Si_3N_4; graphene; lubrication; friction; temperature rise

1. Introduction

With the development of advanced manufacturing technology, the devices are often operated under conditions such as high temperature, high pressure, or less lubrication etc. However, traditional metal materials and metal tribo-pairs were not available due to its vulnerability to rupture and corrosion damage. Engineering ceramic tribo-pairs, for example Si_3N_4/Si_3N_4, Al_2O_3/Al_2O_3, etc., have excellent properties such as low density, significant thermal stability, and high hardness [1]. These properties are suitable for a wide variety of tribological applications [2,3]. Therefore, the research and practical application of Si_3N_4/Si_3N_4 as friction and wear components has become one of the hot spots of devices and material science.

Lubrication is an important measure to reduce wear, save energy, and improve industrial efficiency and reliability. In addition, lubricating oil additives are also important for improving the performance of lubricating oil. Nanomaterials as a kind of lubricant additive play a good role in anti-wear and anti-friction. Among them, as the component part of graphite used as the traditional solid lubricating material, graphene has attracted much attention in the tribological field due to its unique friction and wear properties [4,5]. Graphene possesses an extremely thin laminated structure, high load-bearing capacity, high chemical stability, and low surface energy, and thus, it can offer lower shear stress and prevent direct contact between metal interfaces [6]. Liu et al. [7] prepared novel composite coatings of diamond-like carbon/ionic liquid/graphene. They found that 0.075 mg·mL^{-1} graphene in the composite coatings exhibited the lowest friction coefficient, and the highest bearing capacity in a simulated space environment. Huang et al. [8] investigated the tribological behavior of the graphite nanosheets as an additive in paraffin oil. Their result showed that the load-carrying capacity and anti-wear ability of the lubricating oil were improved. Lin et al. [9], likewise, carried out a new kind of lubricating oil containing modified graphene platelets. The results indicated that it could clearly improve the wear resistance and load-carrying capacity of the machine.

Although many studies have shown that graphene as a lubricant additive can improve the tribological properties of sliding trio-pairs, most of the studies have focused on steel/steel tribo-pairs. In our previous study, we investigated the tribological behaviors of Si_3N_4/GCr15 sliding pairs lubricated with graphene oxide [10]. However, there is less attention on the lubrication of Si_3N_4/Si_3N_4 tribo-pairs, as the structure and performance characteristics of engineering ceramic device material are very different from those of metal materials. In addition, there are also few reports on the cooling effect of graphene as a lubricant additive. The purpose of this work is to explore the lubrication and cooling effect for Si_3N_4/Si_3N_4 tribo-pairs by adding graphene to base lubricating oil. Factors influencing friction coefficient include rotational speed, load, and graphene concentration. We have investigated the tribological behavior of Si_3N_4/Si_3N_4 sliding pairs under three different experimental conditions by using a Rtec MFT 5000 Tribometer with the ball-on-disk mode. Finally, the effect on lubrication and cooling and the lubrication mechanism is discussed.

2. Materials and Methods

2.1. Materials

Few-layer graphene (FLG) was obtained from Detong Nanotechnology Co. Ltd. (Qingdao, China). Mobile DTE oil light (Mdol) were selected as the base lubricating oil and purchased from Huijie Development Co. Ltd. (Changsha, China). FLG was obtained by physical methods and used directly without further purification. The kinematic viscosity of Mdol was 5.34 mm^2·s^{-1}, when the temperature was 100 °C and 29.77 mm^2·s^{-1} with a temperature of 40 °C. Mdol containing different concentrations of FLG (0.025, 0.05, 0.075, and 0.1 wt%) was prepared and followed by ultrasonication for about 30 min to make sure that FLG was evenly dispersed in the base lubricating oil.

2.2. Tribological Test

Si_3N_4 ceramic balls and disks were purchased from Zhihai Bearing Co., Ltd. (Shanghai, China). The ball was a commercial product with a diameter of 9.525 mm and a surface roughness of no more than 140 nm. The thickness of the disk was 5 mm and its surface roughness did not exceed 250 nm. The tribological behaviors of Si_3N_4/Si_3N_4 tribo-pairs lubricated without and with FLG were examined using a Rtec MFT5000 Tribometer (Rtec, San Jose, CA, USA) with the ball-on-disk mode, with the cylindrical upper tribo-pairs as a Si_3N_4 ball and the lower tribo-pairs as a Si_3N_4 disk. The temperature of the Si_3N_4/Si_3N_4 sliding pairs was recorded by a temperature-rise detection device. The temperature-rise detection device included a data acquisition device, a standard rod, and the spindle error analyzer software. The temperature sensor used in data acquisition was a magnetic thermistor attached to the cylindrical pin above the Si_3N_4 ceramic ball. It recorded temperature changes in real time during the experiment.

In a typical test, Mdol containing 0.025, 0.05, 0.075, and 0.1 wt% FLG were prepared and ultrasonic for about 30 min to ensure uniform dispersion of the graphene in the base lubricating oil. In order to avoid interference from other factors, the graphene additive lubricating oil did not use a surfactant.

Since factors influencing the tribological properties was carried out, including FLG concentration (0.025, 0.05, 0.075, and 0.1 wt%), load (40, 80, 140 N), and rotational speed (500, 3000 r·min^{-1}). All tribological tests were repeated at least three times. The friction coefficient was automatically recorded by the experimental device and the real-time temperature was recorded by the temperature-rise detection system.

2.3. Characterization

The structure of FLG was imaged with Raman spectroscope (HR800, Horiba, Paris, France) with a confocal Raman microscope mode and a laser wavelength of 532 nm. Nanoparticle analyzer is an instrument that uses a physical method to test the size and distribution of particles. The particle diameter distribution of FLG was measured by the NanoPlus-3 nanoparticle analyzer (Micromeritics, New York, NY, USA). The wear mark depth of the Si_3N_4 disk was characterized by the OLS4100 3D laser measuring microscope (Olympus, Tokyo, Japan). S-4800 scanning electron microscope (SEM, Hitachi, Tokyo, Japan) equipped with an energy-dispersive X-ray spectroscope (EDS, Hitachi, Tokyo, Japan) and Raman spectroscope (Horiba, Paris, France) were used to observe the wear mark of the Si_3N_4 disk.

3. Results and Discussion

3.1. Materials Characterization

Figure 1 illustrates the characterization of FLG. Figure 1a is the SEM image of FLG, showing that FLG retains its original laminated structure, which is transparent with folding at the edges, suggesting very few layers. Figure 1b portrays the Raman spectrum of FLG. Three typical features of FLG, the D-band (1342 cm^{-1}), G-band (1575 cm^{-1}), and 2D-band (2701 cm^{-1}), are observed. As shown in Figure 1b, the peak shape of the 2D-peak is widest, the intensity of the G-peak is very strong, and the intensity ratio of the G to 2D band (I_G/I_{2D}) is more than 1, demonstrating that the graphene samples exhibit a few-layered structure [11,12], as is proved in SEM (Figure 1a). However, the presence of the D-band (1342 cm^{-1}) illustrates the occurrence of disorder and defects in the graphene samples [13,14].

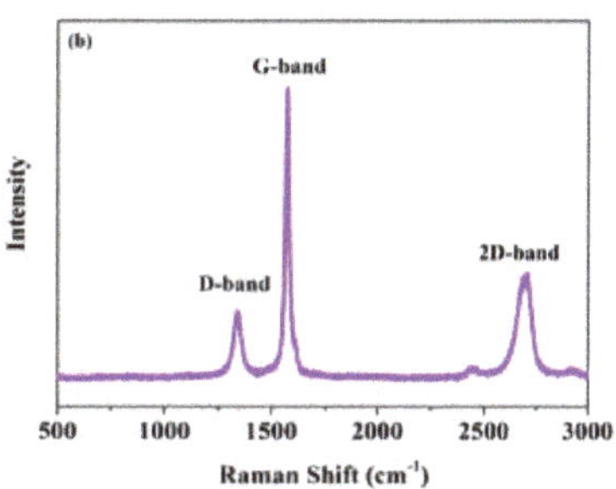

Figure 1. Structure images of few-layer graphene (FLG). (**a**) Scanning electron microscope (SEM) image; (**b**) Raman spectrum.

3.2. Tribological Properties

The average friction coefficient (COF) values of the Si_3N_4/Si_3N_4 sliding pairs lubricated by Mdol with and without FLG at different experimental conditions are shown in Figure 2. Figure 2a shows the friction coefficient curves of the Si_3N_4/Si_3N_4 lubricated by Mdol with 0.1 wt% FLG at 500 r·min^{-1} speed under 40 N loads. When FLG is added to the base oil, the friction coefficient of the Si_3N_4/Si_3N_4 sliding pairs is drastically reduced. Figure 2b shows the COF of the Si_3N_4/Si_3N_4 sliding pairs lubricated with and without different contents of FLG in three different working conditions. Figure 2b clearly shows that different amounts of graphene added to the lubricating oil have less influence on the lubrication effect at low speeds (such as 500 r·min^{-1}), independent of load. Moreover, the lubricating effect is better at high speed and low load. At low speed and high load, the excessive load causes the protective film to rupture. Higher rotational speeds produce a lower COF. When the FLG content is 0.05 wt%, the COF is reduced the least, since the FLG is quickly removed from the wear track due to centrifugal force. As FLG content in base oil is increased, although the FLG is removed by centrifugal force at the same rate, the additional FLG on the wear track improves the lubricating effect depicted by a low friction coefficient. When FLG content is 0.1 wt%, the COF is reduced the most, decreasing by 76.33% (from 0.169 to 0.04).

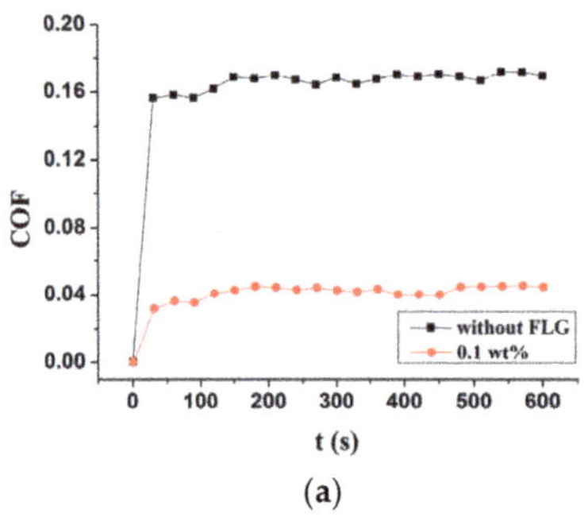

(a)

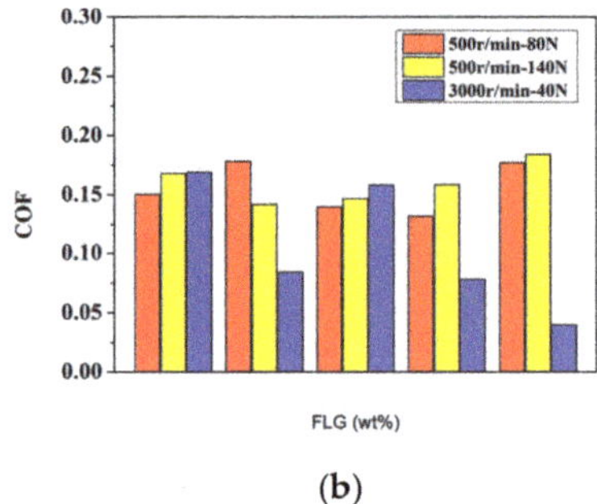

(b)

Figure 2. Friction coefficient (COF) results for Si_3N_4/Si_3N_4 sliding pairs under three working conditions. (**a**) COF with a 40 N load and 3000 r·min^{-1} rotating speed with and without 0.1 wt% FLG; (**b**) comparison of COF results under three different working conditions.

The wear mark depth (WMD) profile curves of the Si_3N_4 disk lubricated by different FLG concentrations under a load of 40 N, at a rotational speed of 3000 r·min^{-1} are shown in Figure 3. It can be seen that the FLG concentration has a certain influence on the WMD of the Si_3N_4 disk. The WMD can be reduced when the Si_3N_4 disk is lubricated by a small amount of FLG. However, when the FLG concentration is too much, it will lead to aggregation and to the increased wear.

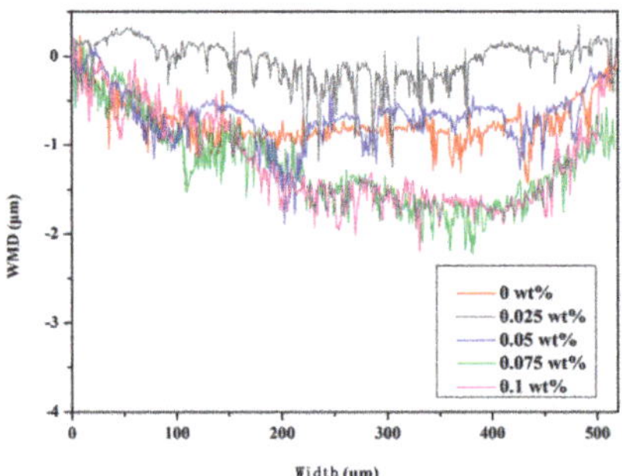

Figure 3. Wear mark depth (WMD) profile curve of the Si_3N_4 disk lubricated by different FLG contents.

Figure 4 shows SEM images of wear scars on the Si_3N_4 disks lubricated by pure Mdol (Figure 4a) and Mdol containing 0.1 wt% FLG (Figure 4b). Many pores and deep scratches can be observed on the rubbing surface lubricated by pure Mdol, and EDS analysis reveals a surface carbon content of only 10.3 wt%. On the contrary, the tribo-surface lubricated by Mdol containing 0.1 wt% FLG becomes significantly smoother with less scratches and fewer pores compared to that lubricated with pure Mdol. In addition, carbon content has increased to more than 25%. This is because FLG enters the interface of tribo-pairs and forms a lubricating protective film, preventing direct contact between the sliding pairs [15].

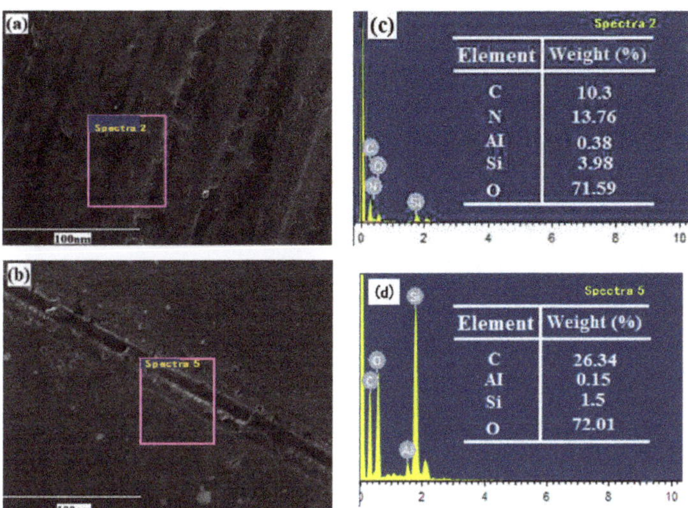

Figure 4. SEM image of worn surfaces lubricated by (**a**) pure Mdol and (**b**), Mdol containing 0.1 wt% FLG; (**c**,**d**) Energy-dispersive X-ray spectroscope (EDS) maps of the areas pointed out in (**a**,**b**).

3.3. Cooling Properties

Friction causes heat and temperature fluctuations during friction and wear testing. Temperature change with time for the Si_3N_4/Si_3N_4 sliding pairs lubricated by Mdol with various contents of FLG under different conditions are shown in Figure 5.

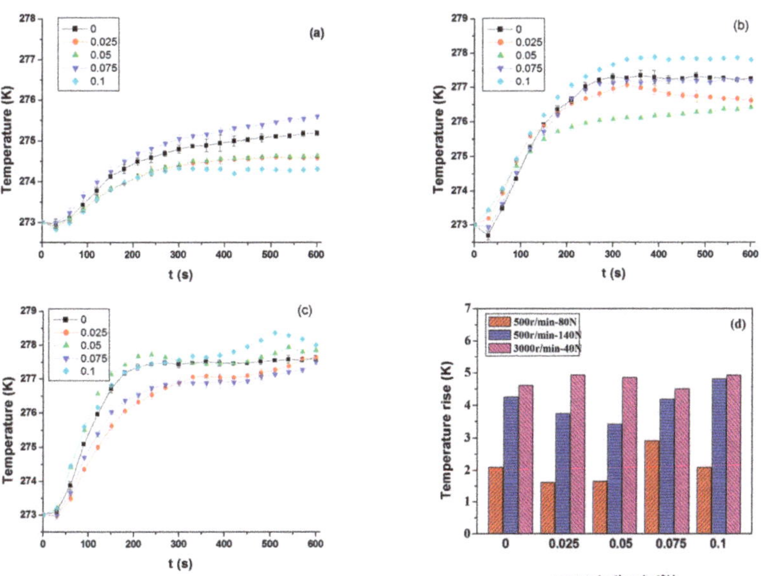

Figure 5. Temperature rise for Si_3N_4/Si_3N_4 sliding pairs under different conditions. (**a**) 80 N load and 500 r·min^{-1} rotational speed, (**b**) 140 N load and 500 r·min^{-1} rotational speed, (**c**) 40 N load and 3000 r·min^{-1} rotational speed, and (**d**) comparison of temperature rise results of the above.

By testing different contents of FLG to the base oil, we observe that a low content of FLG is better in deterring temperature rise at low-speed (500 r·min^{-1} rotating speed). First of all, this can be attributed to the excellent physical properties of FLG (high thermal conductivity and large specific surface area) [16]. When a small amount of FLG is added to Mdol, the overall thermal conductivity of the mixed liquid is increased [17]. During the sliding process, the COF was small, and the lubricant oil flow will take away some of the heat, furthermore, FLG particles distributed in the Mdol will promote heat transfer [18]. Therefore, the heat transfer performance is improved, resulting in a smaller temperature rise. When FLG content increases, a lot of graphene will agglomerate, leading to slower lubricant flow and reduced heat transfer. Secondly, for low rotational speeds, at the same content of FLG, the temperature rise at high loads is higher. This is because the friction coefficient is higher at heavy loads, resulting in more heat generation and therefore a rise in temperature rise. Interestingly, lubricating oils with different graphene concentrations have less influence on the temperature rise at high speed and low load, which is because the centrifugal force has a large influence at high rotation speed, causing some graphene to be scooped out of the wear track. In addition, the friction generates more heat at high speed, which cannot be offset by the cooling effect of a small amount of graphene.

Since Raman spectroscopy is a superior, sensitive, and widely used non-destructive characterization technique and is also an effective tool for assessing the quality of carbon materials. In this work, we use this technique to analyze the rubbing surfaces. Figure 6 shows the Raman spectra of the wear surfaces of the Si_3N_4 disks lubricated by Mdol containing different contents of FLG, as well as that of original FLG powder and in the case of dry friction. In contrast to dry friction (Figure 6a), it can be inferred that the band at 2194 cm^{-1} is the typical signal of Mdol (Figure 6b). When Si_3N_4 disks are lubricated by Mdol containing 0.1 wt% FLG (Figure 6c), Raman spectra of the wear surfaces not only show the lubricating oil signal, but also the FLG signal at the bands 1343 and 1595 cm^{-1}. This indicates that FLG is adhered to the wear surfaces and forms a lubricating protective film. Compared to the Raman signal of FLG powder (Figure 6e), for all FLG (Figure 6d) and Mdol with 0.1% FLG, the intensity ratios of the D and G bands (I_D/I_G) are increased, which illustrates that the adhered graphene has become severely disordered [19], and as a result, the peak shape of 2D becomes weak and wide [20]. Raman spectroscopy was carried out to further confirm that the improved tribological behavior of Si_3N_4/Si_3N_4 sliding pairs is indeed attributable to the presence of FLG on the rubbing surface. These results validate our assumption that FLG is indeed present and has formed a lubricating protective film on the wear surface.

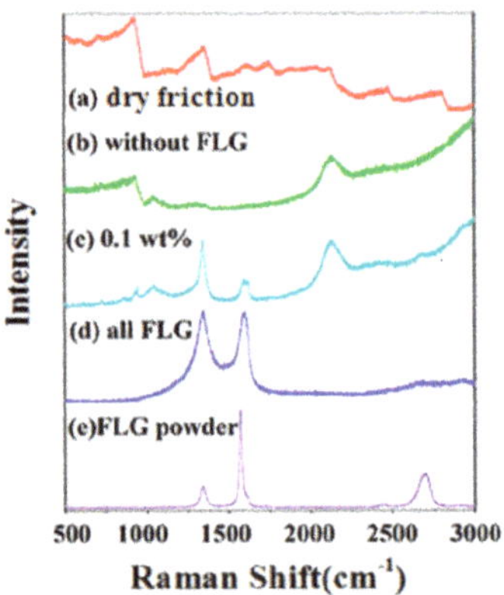

Figure 6. Raman spectra of the rubbing surfaces of the Si_3N_4 disks. (**a**) Without any lubrication; (**b–d**) lubricated by Mdol containing different contents of FLG; (**e**) FLG powder.

Based on the above analysis, a schematic illustration of the tribological mechanism using FLG as a lubricant additive is shown in Figure 7. A schematic diagram of the friction experiment is shown in

Figure 7a. As shown in Figure 7b, direct contact between the Si_3N_4 ceramic ball and disk results in high Hertz contact stress. Therefore, the oil film is quickly broken, and then, the contact surface of Si_3N_4/Si_3N_4 sliding pairs is seriously destroyed. When FLG is added to the base oil, it easily enters the rubbing interface and is sheared owing to its two-dimensional structure [21], as shown in Figure 7c. The addition of FLG into Mdol prevents direct contact between the Si_3N_4/Si_3N_4 sliding pairs because of the formation of a lubricating protective film. In addition, since FLG is easily torn under high Hertz contact pressure, disordered FLG with a much smaller particle size is formed [22]. FLG then wraps the abrasions to prevent direct contact between the sliding pairs. The reduction of the COF and anti-wear properties may be related to these mechanisms. A lubricating protective film was also confirmed in this paper, as measured by EDS analysis and Raman spectroscopy. The lubricating protective film effectively avoids direct contact between sliding pairs, thereby reducing the wear of the surface and decreasing the average friction coefficient.

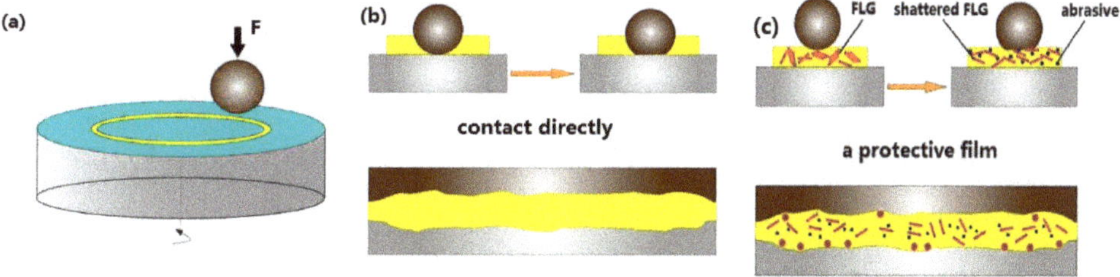

Figure 7. Schematic mechanism for FLG in Mdol. (**a**) Schematic diagram of the friction experiment, (**b**) lubrication diagram of pure lubricating oil, and (**c**) lubrication diagram of FLG suspended in base oil.

4. Conclusions

The effects of FLG as a potential lubricating additive have been studied in detail in different experimental conditions. The results clearly show that FLG has a good effect on Mdol in improving the tribological properties of Si_3N_4/Si_3N_4 sliding pairs for high speeds and low loads, as well as decreasing temperature rise for low speeds regardless of the load level. The FLG content in Mdol has little effect on the width of the disks in the same conditions. On the contrary, it can reduce the WMD of the Si_3N_4 disks. This paper presented a lubrication mechanism to explain the test results. The presence of lubricating protective film can effectively avoid the direct contact between the sliding pairs. In addition, abrasive particles are coated with FLG. As a result, it reduced damage to the surface of the wear scar. FLG as an additive in lubricating oil can provide a good cooling effect for the Si_3N_4/Si_3N_4 sliding pairs, which can be attributed to the high thermal conductivity of FLG.

Author Contributions: Characterization, J.W. and D.X.; Formal analysis, X.W.; Writing—original draft preparation, L.Z.; Writing—review and editing, X.W.; Visualization, D.A.; Project administration, Y.W.; Funding acquisition, L.Z. All authors have read and agreed to the published version of the manuscript.

Funding: This work was supported by the National Science Foundation of China (Nos. 51675353; 51805336), Natural Science Foundation of Liaoning Province (Nos. 20180550002 and 2019-ZD-0687).

Conflicts of Interest: The authors declare no conflict of interest.

References

1. Wang, W.; Hadfield, M.; Wereszczak, A.A. Surface strength of silicon nitride in relation to rolling contact performance. *Ceram. Int.* **2009**, *35*, 3339–3346. [CrossRef]
2. Gates, R.S.; Hsu, S.M. Silicon Nitride Boundary Lubrication: Lubrication Mechanism of Alcohols. *Tribol. Trans.* **1995**, *38*, 645–653. [CrossRef]
3. Kalin, M.; Vizintin, J.; Novak, S.; Dražić, G. Wear mechanisms in oil-lubricated and dry fretting of silicon nitride against bearing steel contacts. *Wear* **1997**, *210*, 27–38. [CrossRef]

4. Kasar, A.K.; Menezes, P.L. Synthesis and recent advances in tribological applications of graphene. *Int. J. Adv. Manuf. Technol.* **2018**, *97*, 3999–4019. [CrossRef]
5. Peng, B.; Locascio, M.; Zapol, P.; Li, S.; Mielke, S.L. Measurements of near-ultimate strength for multiwalled carbon nanotubes and irradiation-induced crosslinking improvements. *Nat. Nanotech.* **2008**, *3*, 626–631. [CrossRef] [PubMed]
6. Schwarz, U.D.; Zwörner, O.; Köster, P.; Wiesendanger, R. Quantitative analysis of the frictional properties of solid materials at low loads. I. Carbon compounds. *Phys. Rev. B* **1997**, *56*, 6987–6996. [CrossRef]
7. Liu, X.; Pu, J.; Xue, Q.; Wang, L. Novel DLC/ionic liquid/graphene nanocomposite coatings towards high-vacuum related space applications. *J. Mater. Chem. A* **2013**, *1*, 3797. [CrossRef]
8. Huang, H.; Tu, J.; Gan, L.; Li, C. An investigation on tribological properties of graphite nanosheets as oil additive. *Wear* **2006**, *261*, 140–144. [CrossRef]
9. Lin, J.; Wang, L.; Chen, G. Modification of Graphene Platelets and their Tribological Properties as a Lubricant Additive. *Tribol. Lett.* **2011**, *41*, 209–215. [CrossRef]
10. Zhang, L.; Zhang, X.; Wu, Y.; Wang, J.; Xi, D. Study on the effects of graphene oxide for tribological properties and cooling in lubricating oil. *Mater. Res. Express* **2018**, *5*, 126509. [CrossRef]
11. Pimenta, M.A.; Dresselhaus, G.; Dresselhaus, M.S.; Cançado, L.G.; Jorio, A.; Saito, R. Studying disorder in graphite-based systems by Raman spectroscopy. *Phys. Chem. Chem. Phys.* **2007**, *9*, 1276–1290. [CrossRef]
12. Graf, D.; Molitor, F.; Ensslin, K.; Stampfer, C.; Jungen, A.; Hierold, C.; Wirtz, L. Spatially Resolved Raman Spectroscopy of Single- and Few-Layer Graphene. *Nano Lett.* **2007**, *7*, 238–242. [CrossRef] [PubMed]
13. Thomsen, C.; Reich, S. Double Resonant Raman Scattering in Graphite. *Phys. Rev. Lett.* **2000**, *85*, 5214–5217. [CrossRef]
14. Ferrari, A.C.; Basko, D.M. Raman spectroscopy as a versatile tool for studying the properties of graphene. *Nat. Nanotechnol.* **2013**, *8*, 235–246. [CrossRef] [PubMed]
15. Wu, L.; Xie, Z.; Gu, L.; Song, B.; Wang, L. Investigation of the tribological behavior of graphene oxide nanoplates as lubricant additives for ceramic/steel contact. *Tribol. Int.* **2018**, *128*, 113–120. [CrossRef]
16. Zheng, Q.; Kim, J.-K. (Eds.) Synthesis, structure, and properties of graphene and graphene oxide. In *Graphene for Transparent Conductors: Synthesis, Properties and Applications*; Springer: New York, NY, USA, 2015; pp. 29–94.
17. Balandin, A.A.; Ghosh, S.; Bao, W.; Calizo, I.; Teweldebrhan, D.; Miao, F.; Lau, C.N. Superior Thermal Conductivity of Single-Layer Graphene. *Nano Lett.* **2008**, *8*, 902–907. [CrossRef] [PubMed]
18. Rutkowski, P.; Stobierski, L.; Górny, G. Thermal stability and conductivity of hot-pressed Si_3N_4–graphene composites. *J. Therm. Anal. Calorim.* **2014**, *116*, 321–328. [CrossRef]
19. Song, H.; Wang, Z.; Yang, J. Tribological properties of graphene oxide and carbon spheres as lubricating additives. *Appl. Phys. A* **2016**, *122*, 933. [CrossRef]
20. Wang, Y.; Pu, J.; Xia, L.; Ding, J.; Yuan, N.; Zhu, Y.; Cheng, G. Fabrication and Tribological Study of Graphene Oxide/Multiply-Alkylated Cyclopentanes Multilayer Lubrication Films on Si Substrates. *Tribol. Lett.* **2014**, *53*, 207–214. [CrossRef]
21. Berman, D.; Erdemir, A.; Sumant, A.V. Reduced wear and friction enabled by graphene layers on sliding steel surfaces in dry nitrogen. *Carbon* **2013**, *59*, 167–175. [CrossRef]
22. Wu, L.; Gu, L.; Xie, Z.; Zhang, C.; Song, B. Improved tribological properties of Si_3N_4/GCr_{15} sliding pairs with few layer graphene as oil additives. *Ceram. Int.* **2017**, *43*, 14218–14224. [CrossRef]

Article

Lift-Off Assisted Patterning of Few Layers Graphene

Alessio Verna [1,*], Simone Luigi Marasso [1,2], Paola Rivolo [1], Matteo Parmeggiani [1,3], Marco Laurenti [1] and Matteo Cocuzza [1,2]

1 Chilab—Materials and Microsystems Laboratory, DISAT, Politecnico di Torino—Via Lungo Piazza d'Armi 6, IT 10034 Chivasso (Torino), Italy; simone.marasso@polito.it (S.L.M.); paola.rivolo@polito.it (P.R.); matteo.parmeggiani@polito.it (M.P.); marco.laurenti@polito.it (M.L.); matteo.cocuzza@infm.polito.it (M.C.)
2 CNR-IMEM, Parco Area delle Scienze 37a, IT 43124 Parma, Italy
3 Center for Sustainable Future Technologies, Italian Institute of Technology, Via Livorno 60, IT 10144 Torino, Italy
* Correspondence: alessio.verna@polito.it; Tel.: +39-011-9114899

Received: 29 May 2019; Accepted: 21 June 2019; Published: 25 June 2019

Abstract: Graphene and 2D materials have been exploited in a growing number of applications and the quality of the deposited layer has been found to be a critical issue for the functionality of the developed devices. Particularly, Chemical Vapor Deposition (CVD) of high quality graphene should be preserved without defects also in the subsequent processes of transferring and patterning. In this work, a lift-off assisted patterning process of Few Layer Graphene (FLG) has been developed to obtain a significant simplification of the whole transferring method and a conformal growth on micrometre size features. The process is based on the lift-off of the catalyst seed layer prior to the FLG deposition. Starting from a SiO_2 finished Silicon substrate, a photolithographic step has been carried out to define the micro patterns, then an evaporation of Pt thin film on Al_2O_3 adhesion layer has been performed. Subsequently, the Pt/Al_2O_3 lift-off step has been attained using a dimethyl sulfoxide (DMSO) bath. The FLG was grown directly on the patterned Pt seed layer by Chemical Vapor Deposition (CVD). Raman spectroscopy was applied on the patterned area in order to investigate the quality of the obtained graphene. Following the novel lift-off assisted patterning technique a minimization of the de-wetting phenomenon for temperatures up to 1000 °C was achieved and micropatterns, down to 10 µm, were easily covered with a high quality FLG.

Keywords: graphene; patterning; Pt; 2D materials; chemical vapor deposition (CVD)

1. Introduction

In the last years, Single Layer or Few Layer Graphene (SLG-FLG) have been widely exploited to obtain novel high performance devices for different types of applications from electronics [1] to energy storage [2], sensors [3], biomedical implants [4] and others [5,6]. The quality of the SLG or FLG has been found to be a critical issue for the functionality of the devices and hence it is fundamental to improve the production and synthesis steps to avoid defects. To develop an SLG and FLG based device, the typical technological processes involved are: Chemical Vapor Deposition (CVD) on metal catalyst seed layers such as Cu or Ni foils [6]; the transferring step on polymer,—that is, poly(methyl methacrylate) (PMMA)—by spin coating and wet etching; the deposition on active regions, that is, metallic electrodes on SiO_2 finished Si substrates; and finally, the patterning step by plasma etching, laser ablation or other techniques [7]. The transfer of graphene onto arbitrary substrates is generally accomplished by polymer-assisted procedures. The transfer process consists of removing graphene from the growing substrate with the aid of a sacrificial polymer layer. For this purpose, several polymers like poly(methyl methacrylate) and polyvinylidene fluoride (PVDF) are widely employed as sacrificial supports [7,8]. Removing of the growing catalyst substrate is accomplished by chemical etching [9] or by electrochemical delamination (ED) [10]. Finally, the graphene/polymer self-standing membrane is attached on the target substrate and the polymer is then removed with a suitable solvent [8]. Despite being widely used, this approach may lead to the formation of undesirable defects and damages within the graphene layer. Hence, a major challenge is to minimize such defects by contact transfer processes at wafer scale [11]. As an alternative, the direct growth of graphene on insulating substrates using Cu vapor [12] or ultrathin Cu [13,14] and Ni [15] films as catalyst has been attempted; however, these methods do not allow for a precise control of the defects. Another interesting approach is to pattern the graphene directly on Cu substrates before transfer [16] ensuring a precise reproduction of the pattern but involving time consuming polymer transfer. Due to the difficulties to grow graphene directly on Cu or Ni thin films and due to the incompatibility of these metals in most biological/electrochemical applications, usually graphene is transferred on Au or Pt electrodes exploiting the previously described methods.

Several applications require the use of Pt electrodes coated with graphene. These include atomic force microscope (AFM) tips for nanoscale electrical characterization [17], friction reduction in Micro Electro Mechanical System (MEMS) [18], counter electrodes for dye-sensitized solar cells [19], microelectrodes for neurostimulation [20], amperometric sensors in electrophoresis devices [21], electrodes for the electrochemical adsorption of dyes by cyclic voltammetry [22] and electrodes decorated with Pt nanoparticles for electrochemical applications [23,24]. Furthermore, graphene patterning on this type of electrodes is crucial to define the active area of graphene on devices like sensors [3,25,26]. This step introduces further defects and damages on the edges [27] or needs for additional technological processes [28].

Due to the high melting point (T_m = 1768 °C) and low vapor pressure (1.3×10^{-14} mmHg), Pt is a well-known catalyst for growing graphene by CVD [29]. In fact, it has been demonstrated that sputtered Pt thin films do not suffer from de-wetting issues, typical of Cu and Ni, during monolayer or few layers graphene growth [30]. The stability of Pt thin films against de-wetting [31] phenomena at high temperature on Si/SiO_2 wafers can be also tuned and improved by increasing the thickness of the film [30]. In addition, the use of adhesion layers may lead to similar results with thinner films as well as e-beam evaporated films. Usual adhesion layers for Pt are metals like Ti, Ta or Cr. However, their exposure to high temperatures causes inter-diffusion or oxide formation and their subsequent degradation [32]. Alumina (Al_2O_3) is preferred as Pt adhesion layer in high temperature applications [33] due to its thermal and chemical stability. Moreover, it can be deposited with common techniques such as sputtering or e-beam evaporation [34] and then easily integrated in Pt deposition process. Finally, Pt is also an optimal choice for electrodes in chemical/biological sensors [35] as well as for high temperature micro-hotplates sensors [36,37] due to its high chemical and temperature stability. Therefore, the direct growth of high quality graphene on patterned Pt thin films may represent an advantage and simplification of the entire device fabrication process.

Here, the direct growth of CVD FLG on Pt thin film was obtained by a lift-off assisted patterning. Al_2O_3 was used as adhesion layer to avoid de-wetting of Pt film. FLG was grown on patterned Al_2O_3/Pt substrates with features down to 10 μm. The graphene quality on patterned areas was evaluated and compared to the graphene grown on un-patterned film by Raman analysis.

2. Materials and Methods

2.1. *Lift-Off Assisted Patterning*

The proposed novel method is based on the lift-off of the catalyst seed layer prior to the FLG deposition (Figure 1). Single side polished, P type, (100) silicon wafers (resistivity 1–10 Ω·cm) finished with 1 μm thermal oxide (supplied by Si-Mat, Kaufering, Germany) were employed for the patterning process. Two cm × two cm squares samples were used in order to fit into the graphene deposition system, which was the NANOCVD-8G system from Moorfield Nanotechnology Ltd. (Cheshire, UK). Samples were cleaned in an acetone bath, rinsed with isopropyl alcohol and then patterned using image reversal photoresist (Microchemicals AZ 5214E, Ulm, Germany) and standard UV (ultraviolet) lithography, through a photomask. The next step was the deposition of 30 nm of Al_2O_3 (purity 99.99%) followed by 60 nm of Pt (purity 99.99%) by electron beam evaporation (ULVAC EBX-14D, Chigasaki, Japan) with a deposition rate of 2–3 Å/s, both for Al_2O_3 and Pt, in high vacuum ($<10^{-5}$ mTorr) and heating the samples at 150 °C during the deposition process.

After the deposition, the photoresist was stripped with dimethyl sulfoxide (DMSO) at 50 °C and the samples were rinsed with deionized water (DI) and dried with nitrogen.

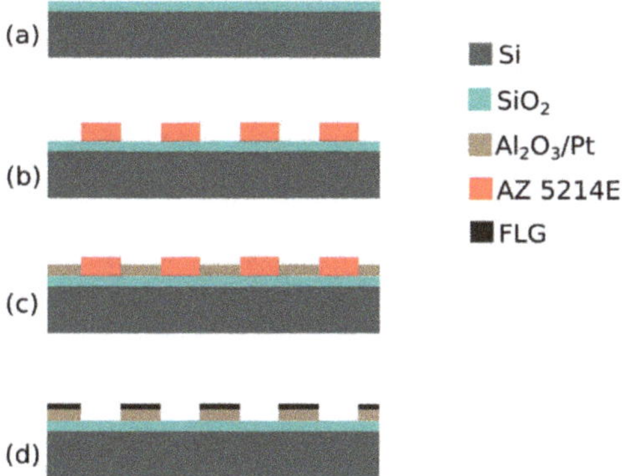

Figure 1. Process flow: (**a**) starting substrate, (**b**) photolithography, (**c**) Al_2O_3/Pt deposition and lift-off, (**d**) graphene growth.

2.2. Graphene Deposition on Pt Film

Graphene was grown on $Si/SiO_2/Al_2O_3/Pt$ substrates by a cold-wall chemical vapor deposition (CVD) reactor operating at low pressure. To remove contaminants from the surface, a two-step annealing of the substrates was performed: 2 min at 900 °C under reducing flow of Ar at 190 sccm and H_2 at 10 sccm (flow control regime) and then 30 s at 1000 °C in an atmosphere composed of Ar 90% and H_2 10% at 10 torr (pressure control regime). The growth of graphene was then carried out at 1000 °C for 300 s, in a mixed atmosphere of Ar (80%), H_2 (10%) and CH_4 (10%) at 10 torr. The samples were finally cooled down to 200 °C under a reducing flow of Ar + H_2 (190 sccm and 10 sccm) and then to room temperature in Ar atmosphere.

2.3. Characterization

Un-patterned Al_2O_3/Pt thin films were characterized with field emission scanning electron microscopy (FESEM) after an annealing treatment at 900 °C, 1000 °C and 1050 °C to investigate the high temperature effect related to the CVD graphene growth process. Images were obtained with FESEM ZEISS Supra 40 (Oberkochen, Germany). For this purpose, the annealing was performed in the same atmosphere and time duration previously described for the growth of graphene but excluding CH_4 in the gas mixture.

X-Ray Diffraction (XRD) was performed on un-patterned $Si/SiO_2/Al_2O_3/Pt$ substrates with the twofold aim of analysing the corresponding crystal structure and orientation and evaluating the effect of the thermal annealing at 1000 °C. XRD patterns were collected using a Panalytical X'Pert Diffractometer (PANalytical, Almelo, The Netherlands) in Bragg-Brentano configuration, equipped with a Cu Kα radiation as X-ray source (λ = 1.54059 Å).

Pt/FLG substrates were characterized by means of a Renishaw InVia Reflex micro-Raman spectrometer (Renishaw plc, Wottonunder-Edge, UK), equipped with a cooled CCD camera. The Raman source was a laser diode (λ = 514.5 nm) and samples inspection occurred in backscattering light collection through a 50× microscope objective for all the single spectra acquisition. The spectra of the patterned structures were obtained by focusing the laser spot on their centre, while a Raman map of the 10 μm-wide circle was collected by scanning, by means of a long working distance 100× objective, a 16 μm × 16 μm area, with a 0.5 μm step. The spectral map analysis was performed by means of the Renishaw WiRE 3.2 software. To collect both the single spectra and the map, 50 mW laser power, 60 s of exposure time and 4 accumulations were employed.

Optical images were acquired with a Nikon Eclipse ME600 microscope (Nikon, Tokyo, Japan).

3. Results

Lift-off assisted patterning of FLG has been successfully obtained (Figure 2) on Al_2O_3/Pt catalyst film.

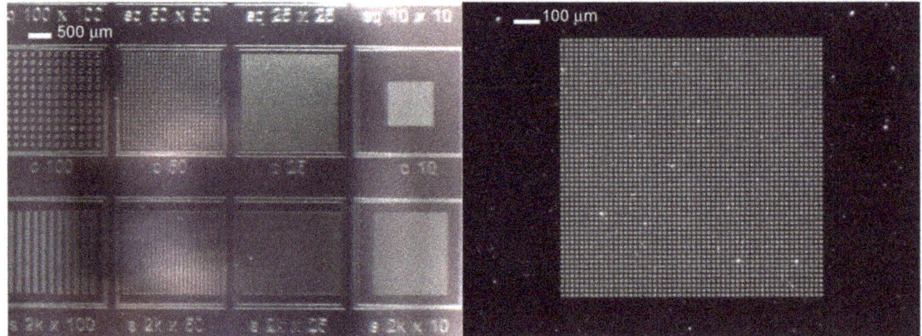

Figure 2. Lift-off assisted patterning of few layers graphene (FLG) on Pt/Al_2O_3 catalyst: optical images of the patterned catalyst with different sizes features.

3.1. Morphologica Characterization of De-Wetting Dynamic

The temperature effects were evaluated by thermal annealing tests on Al_2O_3/Pt layer at 900 °C, 1000 °C and 1050 °C (Figure 3).

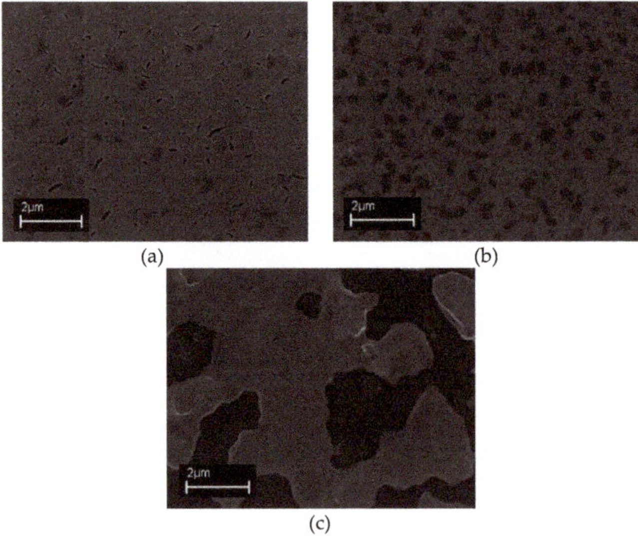

Figure 3. Comparison of field emission scanning electron microscope (FESEM) images of Pt/Al_2O_3 annealed at 900 °C (**a**), 1000 °C (**b**) and 1050 °C (**c**). The film becomes highly discontinuous at 1050 °C although de-wetting process starts below 900 °C.

FESEM images demonstrate that the Al_2O_3 adhesion layer allows for controlling the de-wetting process (see supplementary information) to achieve a FLG growth temperature up to 1000 °C. It can be noticed that a detrimental effect appears at 1050 °C, where a discontinuous film is formed, making it impractical to use for most technological applications.

3.2. XRD Characterization

The diffraction spectrum obtained by XRD investigation (Figure 4) shows a comparison between as-grown Al_2O_3/Pt samples and the 1000 °C annealed one. Apart from the contribution coming from the Si substrate (2θ–69.2°), a single diffraction peak is detected at 40.1° in both cases and ascribed to the family of Pt(111) crystal planes (JCPDS Card 04-0802). After annealing, the crystal quality of the Pt layers turns out to be improved, as demonstrated by the higher Pt(111) peak intensity. Moreover, the (111) crystal orientation is also highly desirable for promoting graphene growth [23]. This characterization validates Al_2O_3 as adhesion layer for this kind of application; indeed, with respect to previous work on a similar process [31], Al_2O_3 prevents a premature de-wetting for e-beam evaporated Pt film.

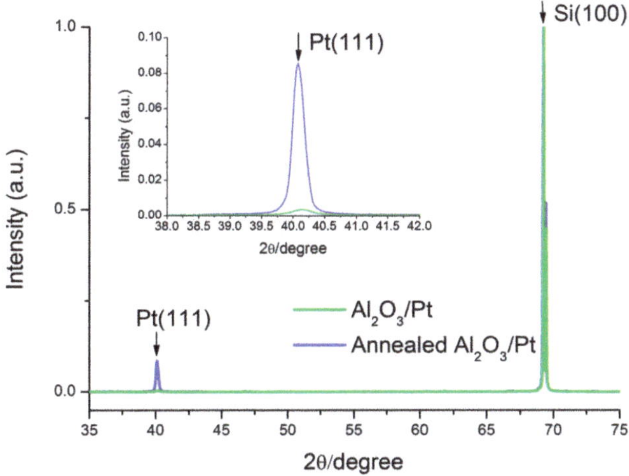

Figure 4. X-ray diffraction (XRD) patterns of Al_2O_3/Pt samples, before and after annealing at 1000 °C. Annealed sample shows the amplification of Pt(111) phase which is suitable for graphene growth. The inset shows a magnification of the Pt(111) phase.

3.3. Raman Characterization of Patterned Pt

The patterned Al_2O_3/Pt was characterized by Raman spectroscopy to evaluate the quality of the grown graphene, which according to Wang et al. was about 2–3 layers [38].

The analysis was performed with the aim to evaluate the selective growth of FLG on Pt patterns with respect to SiO_2 and the correlation between the FLG defectivity and the patterns sizes. Furthermore, growing temperature effect was investigated. Raman spectra of FLG on both un-patterned (red curve) and patterned Pt samples ranging from 5 to 100 μm wide strips were reported (Figure 5). For the patterned Pt samples, the Raman spectra were collected on a 1–2 μm wide area, far from the edges. The intensity (I), position and shape of D, G and 2D peaks (centred at ~1350, 1580 and 2700 cm^{-1} respectively) are similar for the 100 μm patterned and un-patterned areas but the intensity of the D peak differs on the 5 μm pattern. The presence of an ubiquitous peak at ~2324 cm^{-1} can be related to atmospheric N_2 gas fundamental vibration-rotation as previously reported [39].

For the FLG growth on the un-patterned Pt sample, the G peak only differs from the patterned ones in shape and intensity with respect to the D band: a narrower and more symmetric band and an I_D/I_G

ratio of ~0.20 are observable. Moreover, the calculated I_D/I_G is ~0.31 on the 100 µm pattern and ~0.73 on the 5 µm pattern suggesting that the presence of the microstructures could induce a more disordered superposition of the graphene few layers. As pointed out in the review by Ferrari and Basko [40], other factors confirm the disorder induced in the case of patterned Pt/FLG with respect to plain film; these include: the increased dispersion of the G peak; the small elbow at the right of the G peak which could be associated with a small D′ peak and the noise at the left of 2D which could be associated with the D″ peak. On the other hand, the shape and position of the 2D band for both samples (un-patterned and patterned) indicate that the number and the quality of the deposited graphene sheets are quite comparable, as the peak, though symmetric, cannot be fitted by one Lorentzian and it has a FWHM of ~77 cm^{-1}, 64 cm^{-1} and 84 cm^{-1} respectively for un-patterned, 100 µm and 5 µm patterns, which are compatible with the characteristics of FLG grown on a nickel-coated SiO_2/Si substrate, previously reported by Park et al. [41].

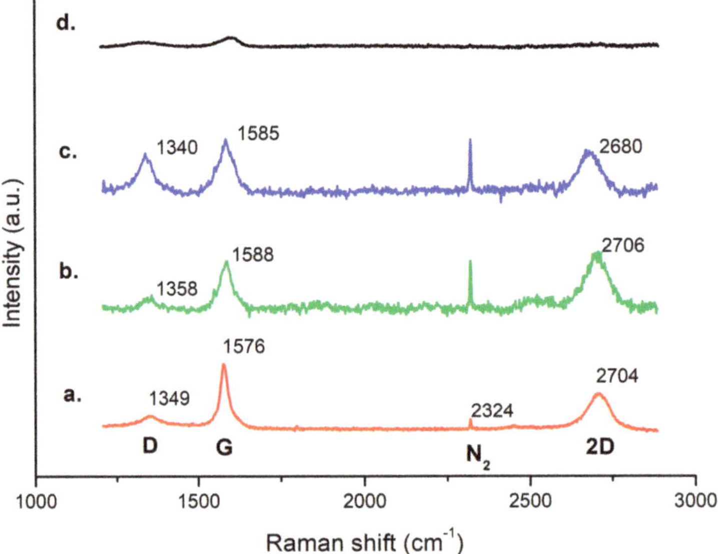

Figure 5. Comparative Raman spectra of un-patterned Pt/FLG (**a**), 100 µm strip pattern (**b**), 5 µm strip pattern (**c**) and blank silicon collected between two 100 µm Pt strips (**d**). The main peaks labels are shown.

Patterned Pt/FLG grown at 900 °C, 1000 °C and 1050 °C was further characterized by Raman spectroscopy (Figure 6). It can be noticed that at 1050 °C graphene quality improves although Pt thin film undergoes de-wetting effect and, with the increasing temperature, the ratio between 2D and G peaks also increases indicating a reduction in the number of layers. Moreover, both D peak intensity reduction and G peak sharpness indicate a minimization of the graphene defects. But from the comparison with morphological analysis (Figure 3), at 1050 °C the de-wetting of the Pt film has relevant detrimental effect.

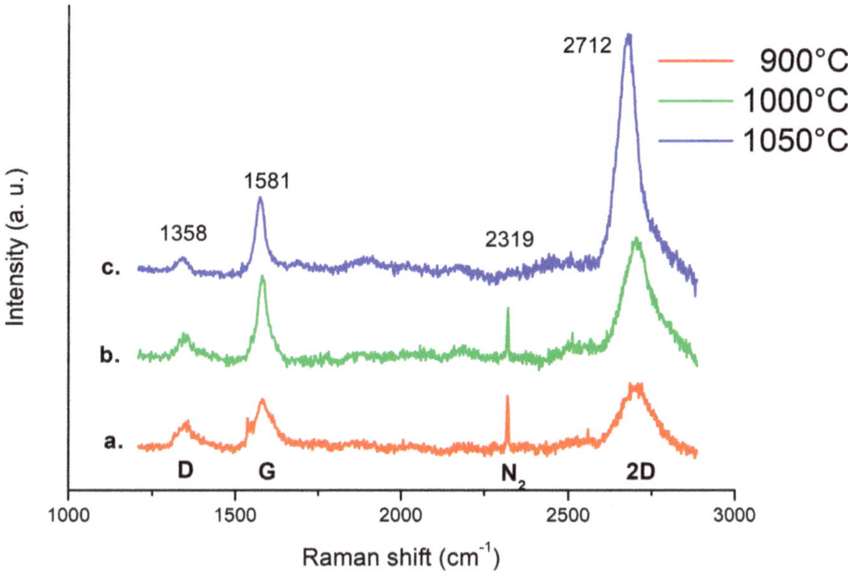

Figure 6. Comparative Raman spectra of 100 μm patterns of Pt/FLG grown at 900 °C (**a**), at 1000 °C (**b**) and at 1050 °C (**c**). The ratio between 2D and G peaks intensities increases indicating a reduction in the number of layers with temperature. The main peaks labels are shown.

Figure 7 shows FLG Raman spectra from the centre of a 5 μm strip to the border of the same pattern and then in a region 2.5 μm far from the edge. A transition from graphene to graphitic carbon residual is observed as the developing of D and G peaks indicate the presence of sp^2 carbon with a consistent number of defects as previously reported [42].

In order to verify the homogeneity of the FLG distribution and possible physical boundary effects, the scanning of a 16 × 16 μm^2 area, including circle-shaped patterns (10 μm in diameter), was performed. The collected spectral Raman map (Figure 8) highlights that, in the inner region of the microstructure, the intensities of the D, G and 2D bands are quite constant in distribution and mutual ratio. Then, all the peak intensities increase by approaching the edge of the micro-circle, suggesting an accumulation of more defective and lower quality graphene sheets within such regions. Beyond the microstructure boundaries, no Raman features related to FLG are present, in accordance with blank spectrum (black curve) of Figure 6. This demonstrates the high selectivity of the growing process. Regarding defects accumulation on the edges, it is possible to assume that the discontinuities on the catalyst can affect the formation of graphene crystal domains thus leading to a more disordered growth.

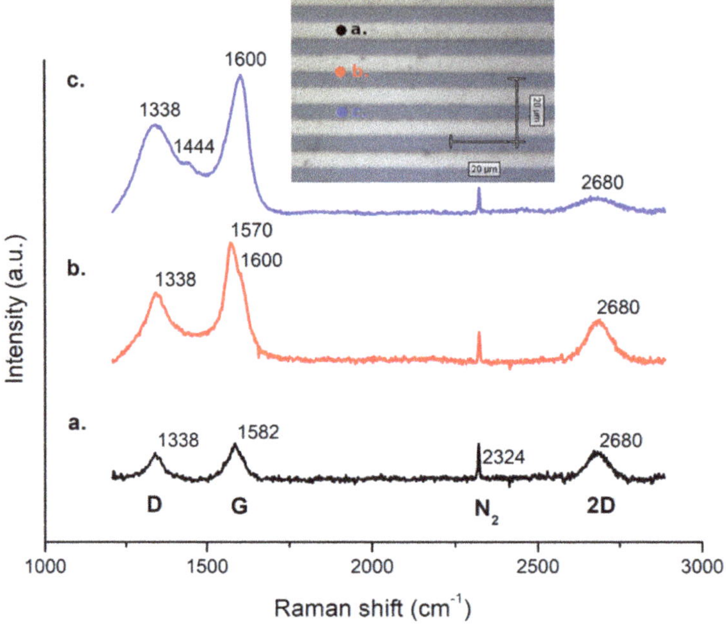

Figure 7. Raman spectra on 5 µm wide strips at different positions at the centre of the strip (**a**), on the border (**b**) and between two strips (**c**). The main peaks labels are shown. The inset shows an optical image of the sample. A graphitic carbon residual is observed in (**c**) while in un-patterned areas (Figure 5d) no carbon residuals are present. This shows that amorphous carbon deposition is catalysed by the presence of platinum also in a halo of 1–2 µm outside the pattern.

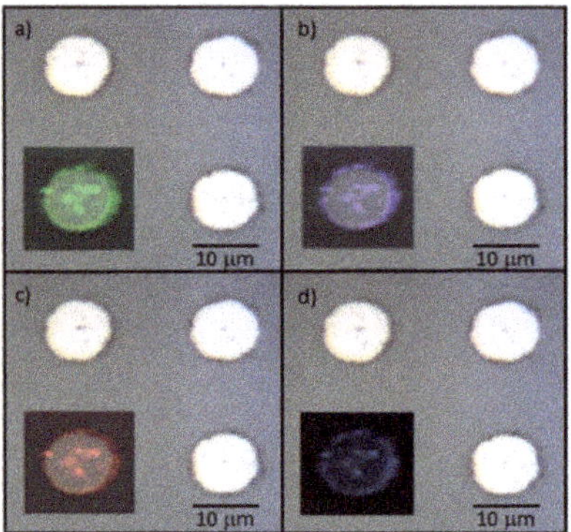

Figure 8. Raman map of a 10 µm wide circle. The map is superimposed to the optical image of 4 circles pattern for comparison. Map shows intensities distribution at 1350 cm^{-1} (**a**), 1580 cm^{-1} (**b**), 2700 cm^{-1} (**c**), overlap of all the selected Raman shift intensity distributions (**d**).

4. Discussion

The reported analysis demonstrates that the lift-off assisted patterning is a valid method to obtain good quality FLG on Pt layer. A significant time reduction with respect to traditional process was achieved since the typical transferring steps were completely skipped. In addition, this method is not affected by the contamination of supporting polymers as PMMA. The obtained optimal repeatability on micrometric patterns allows for covering Pt film with every layouts and, more important, Pt can be deposited with common techniques such as sputtering or e-beam evaporation and then easily integrated in a full device fabrication process. Pt represents an optimal metal selection for electrodes in chemical/biological sensors [5] as well as for high temperature micro-hotplates in Micro Electro Mechanical System (MEMS) [6], due to its high chemical and temperature stability and hence the implementation of this method is of high relevance for a wide range of applications from biosensing to neuronal stimulation.

5. Conclusions

This FLG was grown directly on the patterned Pt seed layer by Chemical Vapor Deposition (CVD). The use of a proper adhesion layer, Al_2O_3, for the Pt film allows for raising the FLG growth temperature up to 1000 °C. The lift-off process of the catalyst, obtained by a standard photolithographic step, leads to a significant time reduction and consequent costs, of the graphene patterning since the typical transferring and etching steps were completely skipped, moreover an optimal repeatability on micrometric patterns can be easily obtained. The Raman characterization shows that the micropatterning was effective, and an accumulation of defects was mostly observed on the edges due to the discontinuity of the patterns. Since Pt is one of the most used materials for electrochemical or gas sensors due to its high thermal and chemical stability, the presented patterning approach has a potential high impact on the fabrication of graphene-based devices, when high quality graphene is required on noble metal electrodes. Moreover, the presented process can be applied to fabricate microelectrodes directly decorated with graphene on a whole wafer of any size avoiding the constraints correlated to polymer-assisted graphene transfer and etching.

Supplementary Materials: The following are available online at http://www.mdpi.com/2072-666X/10/6/426/s1, Figure S1: FESEM images at different magnifications of graphene growth on Pt at 900 °C. Figure S2: FESEM images at different magnifications of graphene growth on Pt at 1000 °C. Figure S3: FESEM images at different magnifications of graphene growth on Pt at 1050 °C. Table S1: Percentage of Pt coverage to evaluate de-wetting.

Author Contributions: Conceptualization, A.V.; validation, P.R., M.L.; formal analysis, P.R. and M.P.; investigation, A.V.; writing—original draft preparation, A.V.; writing—review and editing, S.L.M.; supervision, M.C.

Funding: This research received no external funding.

Conflicts of Interest: The authors declare no conflict of interest.

References

1. Dragoman, M.; Dragoman, D. Graphene-based quantum electronics. *Prog. Quantum Electron.* **2009**, *33*, 165–214. [CrossRef]
2. Marasso, S.L.; Rivolo, P.; Giardi, R.; Mombello, D.; Gigot, A.; Serrapede, M.; Benetto, S.; Enrico, A.; Cocuzza, M.; Tresso, E.; et al. A novel graphene based nanocomposite for application in 3D flexible micro-supercapacitors. *Mater. Res. Express* **2016**, *3*, 065001. [CrossRef]
3. Hill, E.W.; Vijayaragahvan, A.; Novoselov, K. Graphene sensors. *IEEE Sens. J.* **2011**, *11*, 3161–3170. [CrossRef]
4. Podila, R.; Moore, T.; Alexis, F.; Rao, A. Graphene Coatings for Biomedical Implants. *J. Vis. Exp.* **2013**, *73*, 1–9. [CrossRef] [PubMed]
5. Zhang, Y.I.; Zhang, L.; Zhou, C. Review of Chemical Vapor Deposition of Graphene and Related Applications. *Acc. Chem. Res.* **2013**, *46*, 2329–2339. [CrossRef] [PubMed]
6. Ghorban Shiravizadeh, A.; Elahi, S.M.; Sebt, S.A.; Yousefi, R. High performance of visible-NIR broad spectral photocurrent application of monodisperse PbSe nanocubes decorated on rGO sheets. *J. Appl. Phys.* **2018**, *123*, 083102. [CrossRef]
7. Barin, G.B.; Song, Y.; Gimenez, I.D.F.; Filho, A.G.S.; Barreto, L.S.; Kong, J. Optimized graphene transfer: Influence of polymethylmethacrylate (PMMA) layer concentration and baking time on grapheme final performance. *Carbon* **2015**, *84*, 82–90. [CrossRef]
8. Song, J.; Kam, F.; Png, R.; Seah, W.; Zhuo, J.; Lim, G.; Ho, P.K.H.; Chua, L. A general method for transferring graphene onto soft surfaces. *Nat. Nanotechnol.* **2013**, *8*, 356–362. [CrossRef]

9. Lee, Y.; Bae, S.; Jang, H.; Jang, S.; Zhu, S.-E.; Sim, S.H.; Song, Y.I.; Hong, B.H.; Ahn, J.-H. Wafer-Scale Synthesis and Transfer of graphene films. *Nano Lett.* **2010**, *10*, 490–493. [CrossRef]
10. Gao, L.; Ren, W.; Xu, H.; Jin, L.; Wang, Z.; Ma, T.; Ma, L.; Zhang, Z.; Fu, Q.; Peng, L.; et al. Repeated growth and bubbling transfer of graphene with millimetre-size single-crystal grains using platinum. *Nat. Commun.* **2012**, *3*, 699. [CrossRef]
11. Gao, L.; Ni, G.-X.; Liu, Y.; Liu, B.; Castro Neto, A.H.; Loh, K.P. Face-to-face transfer of wafer-scale graphene films. *Nature* **2013**, *505*, 190–194. [CrossRef] [PubMed]
12. Kim, H.; Song, I.; Park, C.; Son, M.; Hong, M.; Kim, Z.; Kim, J.S.; Shin, H.; Baik, J.; Choi, C. Copper-Vapor-Assisted Chemical Vapor Deposition for High-Quality and Metal-Free Single-Layer Graphene on Amorphous SiO_2 Substrate. *ACS Nano* **2013**, *7*, 6575–6582. [CrossRef] [PubMed]
13. Ismach, A.; Druzgalski, C.; Penwell, S.; Schwartzberg, A.; Zheng, M.; Javey, A.; Bokor, J.; Zhang, Y. Direct chemical vapor deposition of graphene on dielectric surfaces. *Nano Lett.* **2010**, *10*, 1542–1548. [CrossRef] [PubMed]
14. Kaplas, T.; Sharma, D.; Svirko, Y. Few-layer graphene synthesis on a dielectric substrate. *Carbon* **2012**, *50*, 1503–1509. [CrossRef]
15. Marchena, M.; Janner, D.; Chen, T.L.; Finazzi, V.; Pruneri, V. Low temperature direct growth of graphene patterns on flexible glass substrates catalysed by a sacrificial ultrathin Ni film. *Opt. Mat. Express* **2016**, *6*, 2487–2507. [CrossRef]
16. Alexeev, A.M.; Barnes, M.D.; Nagareddy, V.K.; Craciun, M.F.; Wright, C.D. A simple process for the fabrication of large-area CVD graphene based devices via selective *in situ* functionalization and patterning. *2D Mater.* **2016**, *4*, 011010. [CrossRef]
17. Lanza, M.; Bayerl, A.; Gao, T.; Porti, M.; Nafria, M.; Jing, G.Y.; Zhang, Y.F.; Liu, Z.F.; Duan, H.L. Graphene-coated atomic force microscope tips for reliable nanoscale electrical characterization. *Adv. Mater.* **2013**, *25*, 1440–1444. [CrossRef]
18. Klemenz, A.; Pastewka, L.; Balakrishna, S.G.; Caron, A.; Bennewitz, R.; Moseler, M. Atomic scale mechanisms of friction reduction and wear protection by graphene. *Nano Lett.* **2014**, *14*, 7145–7152. [CrossRef]
19. Cheng, C.E.; Lin, C.Y.; Shan, C.H.; Tsai, S.Y.; Lin, K.W.; Chang, C.S.; Shih-Sen Chien, F. Platinum-graphene counter electrodes for dye-sensitized solar cells. *J. Appl. Phys.* **2013**, *114*, 014503. [CrossRef]
20. Park, H.; Zhang, S.; Steinman, A.; Chen, Z.; Lee, H. Graphene prevents neurostimulation-induced platinum dissolution in fractal microelectrodes. *2D Mater.* **2019**, *6*, 035037. [CrossRef]
21. Lucca, B.G.; de Lima, F.; Coltro, W.K.T.; Ferreira, V.S. Electrodeposition of reduced graphene oxide on a Pt electrode and its use as amperometric sensor in microchip electrophoresis. *Electrophoresis* **2015**, *36*, 1886–1893. [CrossRef] [PubMed]
22. Molina, J.; Fernández, J.; García, C.; Del Río, A.I.; Bonastre, J.; Cases, F. Electrochemical characterization of electrochemically reduced graphene coatings on platinum. Electrochemical study of dye adsorption. *Electrochim. Acta* **2015**, *166*, 54–63. [CrossRef]
23. Gutés, A.; Hsia, B.; Sussman, A.; Mickelson, W.; Zettl, A.; Carraroa, C.; Maboudian, R. Graphene decoration with metal nanoparticles: Towards easy integration for sensing applications. *Nanoscale* **2012**, *4*, 438–440. [CrossRef] [PubMed]
24. Huang, J.; Tian, J.; Zhao, Y.; Zhao, S. Ag/Au nanoparticles coated graphene electrochemical sensor for ultrasensitive analysis of carcinoembryonic antigen in clinical immunoassay. *Sens. Actuators B Chem.* **2015**, *206*, 570–576. [CrossRef]
25. Shao, Y.; Wang, J.; Wu, H.; Liu, J.; Aksay, I.A.; Lin, Y. Graphene Based Electrochemical Sensors and Biosensors: A Review. *Electroanalysis* **2010**, *22*, 1027–1036. [CrossRef]
26. Yuan, W.; Shi, G. Graphene-based gas sensors. *J. Mater. Chem. A* **2013**, *1*, 10078–10091. [CrossRef]
27. Pérez-Mas, A.M.; Álvarez, P.; Campos, N.; Gómez, D.; Menéndez, R. Graphene patterning by nanosecond laser ablation: The effect of the substrate interaction with graphene. *J. Phys. D. Appl. Phys.* **2016**, *49*, 305301. [CrossRef]
28. Feng, J.; Li, W.; Qian, X.; Qi, J.; Qi, L.; Li, J. Patterning of graphene. *Nanoscale* **2012**, *4*, 4883–4899. [CrossRef] [PubMed]
29. Sutter, P.; Sadowski, J.T.; Sutter, E. Graphene on Pt(111): Growth and substrate interaction. *Phys. Rev. B* **2009**, *80*, 1–10. [CrossRef]

30. Kang, B.J.; Mun, J.H.; Hwang, C.Y.; Cho, B.J. Monolayer graphene growth on sputtered thin film platinum. *J. Appl. Phys.* **2009**, *106*, 104309. [CrossRef]
31. Nam, J.; Kim, D.C.; Yun, H.; Shin, D.H.; Nam, S.; Lee, W.K.; Hwang, J.Y.; Lee, S.W.; Weman, H.; Kim, K.S. Chemical vapor deposition of graphene on platinum: Growth and substrate interaction. *Carbon* **2017**, *111*, 733–740. [CrossRef]
32. Firebaugh, S.L.; Jensen, K.F.; Schmidt, M.A. Investigation of high-temperature degradation of platinum thin films with an in situ resistance measurement apparatus. *J. Microelectromech. Syst.* **1998**, *7*, 128–135. [CrossRef]
33. Guarnieri, V.; Biazi, L.; Marchiori, R.; Lago, A. Platinum metallization for MEMS application: Focus on coating adhesion for biomedical applications. *Biomatter* **2014**, *4*, 2–8. [CrossRef] [PubMed]
34. Madaan, N.; Kanyal, S.S.; Jensen, D.S.; Vail, M.A.; Dadson, A.E.; Engelhard, M.H.; Samha, H.; Linford, M.R. Al_2O_3 e-Beam Evaporated onto Silicon (100)/SiO_2, by XPS. *Surf. Sci. Spectra* **2013**, *20*, 43–48. [CrossRef]
35. Tarabella, G.; Balducci, A.G.; Coppedè, N.; Marasso, S.; D'Angelo, P.; Barbieri, S.; Cocuzza, M.; Colombo, P.; Sonvico, F.; Mosca, R.; et al. Liposome sensing and monitoring by organic electrochemical transistors integrated in microfluidics. *Biochim. Biophys. Acta, Gen. Subj.* **2013**, *1830*, 4374–4380. [CrossRef] [PubMed]
36. Marasso, S.L.; Tommasi, A.; Perrone, D.; Cocuzza, M.; Mosca, R.; Villani, M.; Zappettini, A.; Calestani, D. A new method to integrate ZnO nano- tetrapods on MEMS micro-hotplates for large scale gas sensor production. *Nanotechnology* **2016**, *27*, 1–7. [CrossRef] [PubMed]
37. Tommasi, A.; Cocuzza, M.; Perrone, D.; Pirri, C.; Mosca, R.; Villani, M.; Delmonte, N.; Zappettini, A.; Calestani, D.; Marasso, S. Modeling, Fabrication and Testing of a Customizable Micromachined Hotplate for Sensor Applications. *Sensors* **2016**, *17*, 62. [CrossRef] [PubMed]
38. Wang, Y.; Ni, Z.; Yu, T.; Shen, Z.X.; Wang, H.; Wu, Y.; Chen, W.; Thye, A.; Wee, S. Raman Studies of Monolayer Graphene: The Substrate Effect. *J. Phys. Chem. C* **2008**, *112*, 10637–10640. [CrossRef]
39. Hawaldar, R.; Merino, P.; Correia, M.R.; Bdikin, I.; Grácio, J.; Méndez, J.; Martín-Gago, J.A.; Singh, M.K. Large-area high-throughput synthesis of monolayer graphene sheet by Hot filament thermal chemical vapor deposition. *Sci. Rep.* **2012**, *2*, 2–10. [CrossRef]
40. Ferrari, A.C.; Basko, D.M. Raman spectroscopy as a versatile tool for studying the properties of graphene. *Nat. Nanotechnol.* **2013**, *8*, 235–246. [CrossRef]
41. Park, H.J.; Meyer, J.; Roth, S.; Skákalová, V. Growth and properties of few-layer graphene prepared by chemical vapor deposition. *Carbon* **2010**, *48*, 1088–1094. [CrossRef]
42. Jerng, S.-K.; Seong Yu, D.; Hong Lee, J.; Kim, C.; Yoon, S.; Chun, S.-H. Graphitic carbon growth on crystalline and amorphous oxide substrates using molecular beam epitaxy. *Nanoscale Res. Lett.* **2011**, *6*, 565. [CrossRef] [PubMed]

Article

Fano-Resonance in Hybrid Metal-Graphene Metamaterial and Its Application as Mid-Infrared Plasmonic Sensor

Jianfa Zhang *, **Qilin Hong, Jinglan Zou, Yuwen He, Xiaodong Yuan, Zhihong Zhu and Shiqiao Qin**

College of Advanced Interdisciplinary Studies, National University of Defense Technology, Changsha 410073, China; qlhong95@126.com (Q.H.); jinglanzou@sina.com (J.Z.); 18380597763@163.com (Y.H.); x.d.yuan@163.com (X.Y.); zzhwcx@163.com (Z.Z.); sqqin8@nudt.edu.cn (S.Q.)

* Correspondence: jfzhang85@nudt.edu.cn; Tel.: +86-0731-84573737

Received: 15 February 2020; Accepted: 28 February 2020; Published: 4 March 2020

Abstract: Fano resonances in nanostructures have attracted widespread research interests in the past few years for their potential applications in sensing, switching and nonlinear optics. In this paper, a mid-infrared Fano resonance in a hybrid metal-graphene metamaterial is studied. The hybrid metamaterial consists of a metallic grid enclosing with graphene nanodisks. The Fano resonance arises from the coupling of graphene and metallic plasmonic resonances and it is sharper than plasmonic resonances in pure graphene nanostructures. The resonance strength can be enhanced by increasing the number of graphene layers. The proposed metamaterial can be employed as a high-performance mid-infrared plasmonic sensor with an unprecedented sensitivity of about 7.93 μm/RIU and figure of merit (FOM) of about 158.7.

Keywords: fano resonance; graphene; plasmonic sensor

---

1. Introduction

In the past decade, the so called Fano resonance—a type of resonance originated from the constructive and destructive interference of a narrow discrete resonance with a broad spectral line or continuum—has attracted wide spread research interests in the nanophotonics community [1]. Fano resonances have been observed in various dielectric or plasmonic nanostructures such as metamaterials [2–6], oligomers [7], nanocavities [8–10] and so on [11–13]. The sharp variation of the scattering profile by Fano resonances leads to a variety of applications in optics such as sensing, switching and nonlinear devices. Particularly, Fano resonances in plasmonic nanostructures are sensitive to the changes of local environment and a small perturbation can induce dramatic change of scattering profiles [14]. Thus, Fano-resonant plasmonic structures, in combination with the appropriate chemical and bio-markers, could enable the development of label-free chemical and bio-sensors [15–20].

Traditionally, plasmonic nanostructures are built on noble metals such as silver and gold and they work mainly in the visible and near-infrared (IR) ranges. Meanwhile, graphene has recently been rising as a building block for plasmonic devices in the mid- and far-IR ranges [21–23]. Graphene plasmons show relatively low losses, high spatial confinement and incomparable tunability by chemical or electrostatic doping [24,25], providing an versatile platform for tunable infrared devices [26,27], mid-IR sensing [28–31], photodetectors [32,33] and other applications. Hybrid metal-graphene structures have also been studied [34]. The plasmonic resonances of metallic nanostructures can be employed to enhance light-graphene interactions in the visible and IR ranges [35,36] while graphene provides an ideal material to tune the optical properties of metamaterials [34,37–39]. Due to the different plasmonic properties of metal and graphene, a hybrid metal-graphene structure could be designed to show multi resonances in ultra-broadband spectral ranges from near to mid-IR ranges [40]. Their resonances can also coupling with each other, exhibiting interesting resonant behavior such as Fano resonances [34,41,42].

In this paper, a hybrid metal-graphene metamaterial with a Fano-like resonance in the mid-IR range is proposed. The Fano resonance arises from the coupling of the narrowband plasmonic resonance of the graphene nanostructure to the broadband resonance of the metallic structure. Its linewidth is narrower than that of the graphene plasmonic resonance and the resonance strength can be controlled by changing the number of graphene layers. Its potential as a mid-IR refractive index sensor is studied.

2. Results and Discussion

Figure 1a shows the schematic of the hybrid metamaterial. The structure comprises a cover layer, a metallic grid (gold film with periodic holes) enclosing with graphene nanodisks right in its middle and a semi-infinite substrate. The graphene nanodisks are assumed to locate right in the middle of the holes (at the same height of the top surface of the metallic grid). The period of a unit cell is $P = 800$ nm and length of the square hollow in the gold film is $d = 300$ nm. The height of the gold layer is $H = 30$ nm and the diameter of graphene nanodisks is $d = 300$ nm. The structure is excited by a x-polarized wave at normal incidence.

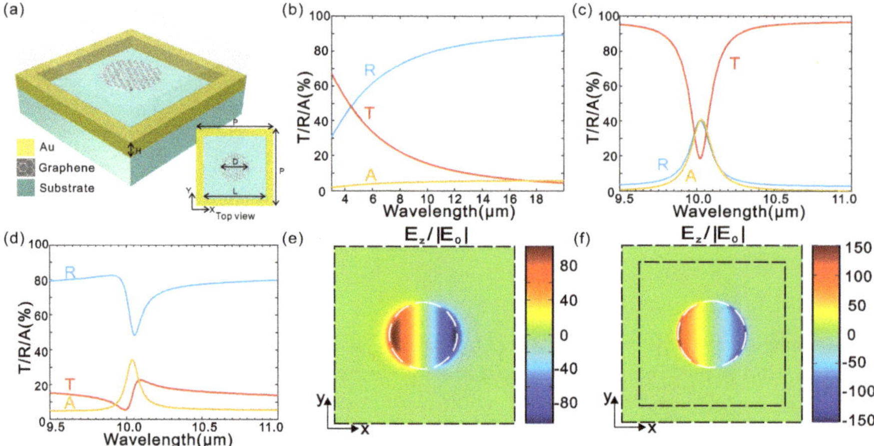

Figure 1. A hybrid metal-graphene Fano-resonant metamaterial. (**a**) Schematic of the hybrid metamaterial. (**b–d**) Simulated spectra of transmission, reflection and absorption for different combination of nanostructured gold film and graphene nanodisks including nanostructured gold film without graphene nanodisks (**b**), graphene nanodisks without the nanostructured gold film (**c**) and nanostructured gold film enclosing with graphene nanodisks (hybrid metamaterial) (**d**). (**e**,**f**) Distributions of local electric fields in the z-direction at the resonance wavelength for graphene nanodisks at ~10 μm and the proposed hybrid metamaterial at ~10.05 μm, respectively. The fields are normalized to the field amplitude of the incident wave (E_0) and plotted at the x-y plane that is 5 nm above the graphene nanodisks. The x-polarized light impinges on the top side of the structure at normal incidence.

For the metallic structure without graphene, the calculated spectra are shown in Figure 1b. The plasmonic resonance is broadband and the resonance peak in the short wavelength range is not shown here. For the periodical array of graphene nanodisks alone, the optical spectra are shown in Figure 1c. There is a plasmonic dipolar resonance (see Figure 1e) at the wavelength of around 10 μm and the full width at half maximum (FWHM) of the resonance is about 0.16 μm. For the hybrid metal-graphene metamaterial, the coupling between the broadband plasmonic resonance of the metallic nanostructure and the narrowband plasmonic dipolar resonance of the graphene nanodisks leads to a sharp Fano resonance at around 10.04 μm with a FWHM of about 0.05 μm (Figure 1d,f).

An effective way to manipulate the mid-IR resonance is controlling the Fermi energy of graphene. However, it is technically challenging to realize Fermi engergy higher than 1 eV. Another way to enhance the plasmonic responses of graphene is using stacked graphene instead of monolayer. Previous studies have shown that infrared plasmonic response of a graphene multi-layer stack is analogous to that of a highly doped single layer of graphene [43]. For simplification, we assume that multilayers of graphene are stacked without separation and it can be replaced by an equivalent layer having the sum of the conductivity of each layer. Figure 2 shows the simulated transmission spectra of the hybrid metal-graphene metamaterial with different layers of graphene. With the increase of graphene layers, the resonance blue shifts and the intensity increases. As the graphene layer increases from 1 to 2, the Fano resonance blue shifts from 10.04 μm to 7.17 μm and amplitude of resonance in transmission increase from 17.3% to 41.1%. Furthermore, the Fano resonance blue shifts to 5.91 μm and amplitude of resonance in transmission increases to 56.3% for 3 layers of graphene.

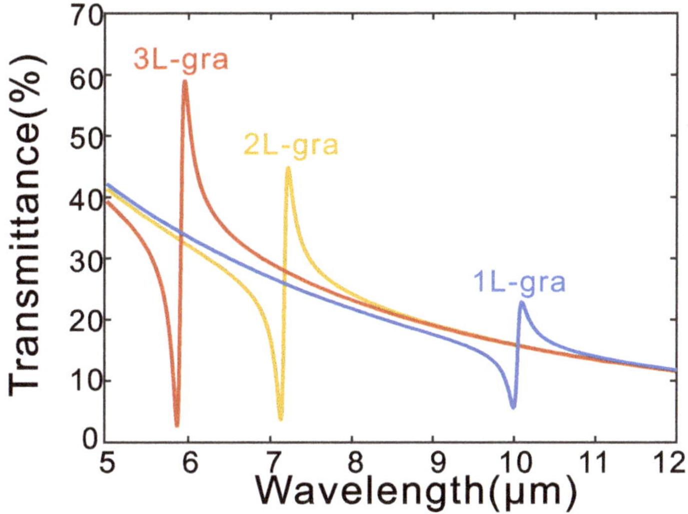

Figure 2. Simulated spectra of transmission for (**a**) Graphene nanodisks and (**b**) The hybrid metal-graphene metamaterial with different layers of graphene.

The sharp Fano resonance of our proposed hybrid graphene-metal metamaterial can increase the figure of merit (FOM, the sensitivity value divided by FWHM) of sensors and is suitable for mid-IR plamsonic sensing. To evaluate its sensing ability, we calculate the transmission spectra of the hybrid metamaterial when it is covered by a semi-infinite layer with different values of refractive indices (Figure 3). As the refractive index of the cover layer (including the medium in the holes of the gold film) increases from 1 to 1.3, the resonance wavelength red shifts from 10.04 μm to 12.42 μm, corresponding to a linear sensitivity of about 7.93 μm/RIU (Figure 3b) and FOM of about 158.7.

In the above, we have shown that the hybrid graphene-metal metamaterial exhibits sharp Fano resonance in the mid-IR range and it can be employed for high-performance sensing. The discussed structure possess several drawbacks in practical realization. First, the graphene nanodisks are electrically isolated and this makes it difficult for electrostatic doping and dynamic modulation of Fermic energy (electrostatic doping with ion-gel is possible but the cover layer will affect its sensing applications). So we may need to employ chemically doped graphene. Secondly, we have assumed that the graphene film is located at the same height of the top surface of the metallic grid which is difficult to fabricate. In order to solve these problems, modified structures can be employed. As an example, we show a hybrid metal-graphene metamaterial where graphene is located directly on the surface of the substrate and the graphene is electrically connected (Figure 4a). Such a structure can be fabricated with standard nanofabrication technique where CVD grown graphene can be transferred to the substrate and patterned by electron beam lithography (EBL). Then the metallic nanostructure can be fabricated on graphene by aligned EBL along with a lift-off process. The simulated spectra of transmission for the hybrid metamaterial with two layers of graphene are shown in Figure 4b. Similar to the spectra in Figure 2, there is a sharp Fano resonance at around 6.60 µm. Besides, an additional resonance appears at around 8.81 µm. These two resonances arise from the coupling between the broadband plasmonic resonance of the metallic nanostructure and the narrowband plasmonic resonance of the graphene nanodisks and that of the graphene cross arms (see insets of field distributions in Figure 4).

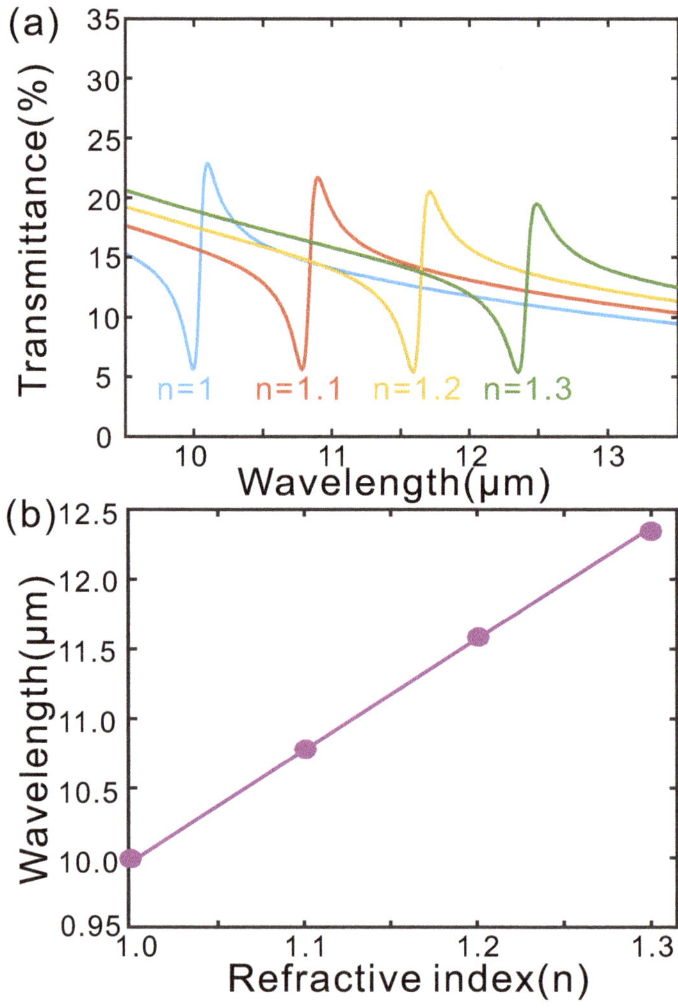

Figure 3. (**a**) Calculated transmittance of the hybrid metal-graphene metamaterial with a cover layer of different refractive indices. (**b**) Wavelengths of the transmittance dips as a function of the cover layer's refractive index.

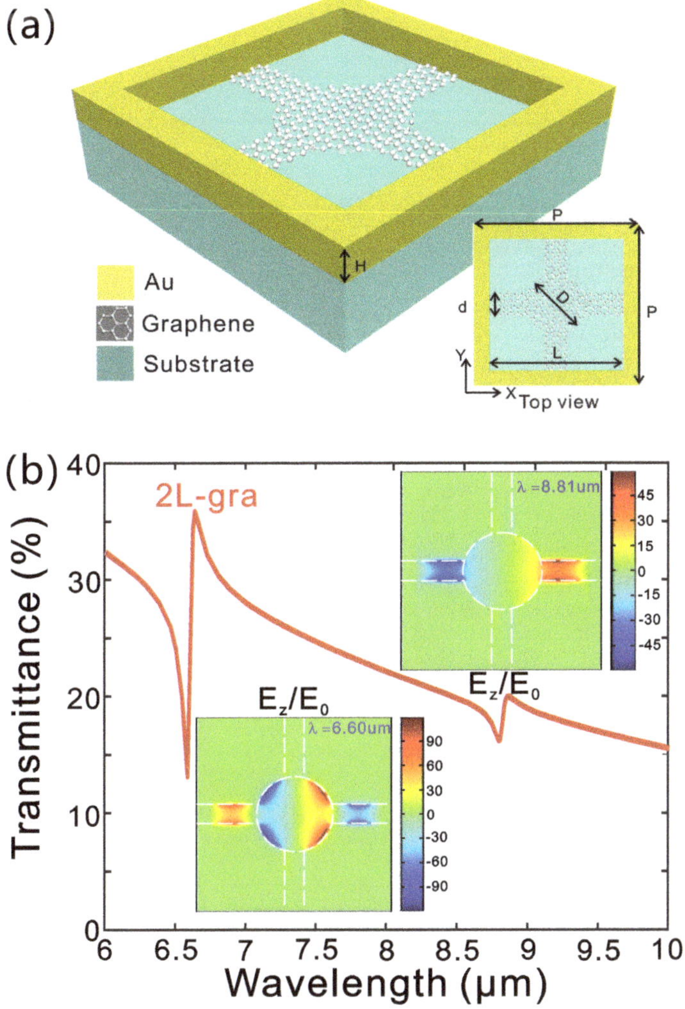

Figure 4. (**a**) Schematic of a modified hybrid metal-graphene metamaterial. (**b**) Simulated spectra of transmission. The insets are field distributions at the two resonances which are normalized to the field amplitude of the incident wave (E_0) and plotted at the x-y plane that is 5 nm below the graphene nanodisks.

3. Materials and Methods

The numerical simulations are conducted using a fully three-dimensional finite element technique (COMSOL Multiphysics, Stockholm, Sweden). In simulations, the monolayer graphene sheet is modeled as a conductive surface [44,45] and the transition boundary is used for it. Optical conductivity of graphene can be derived within the random-phase approximation (RPA) in the local limit as below [46,47].

$$
\begin{aligned}
\sigma(\omega) =& \frac{2e^2 k_B T}{\pi\hbar^2}\frac{i}{\omega + i\tau^{-1}}\ln[2\cosh\frac{E_f}{2k_B T}] + \\
&\frac{e^2}{4\hbar}[\frac{1}{2} + \frac{1}{\pi}\arctan(\frac{\hbar\omega - 2E_f}{2k_B T}) - \\
&\frac{i}{2\pi}\ln\frac{(\hbar\omega + 2E_f)^2}{(\hbar\omega - 2E_f)^2 + (2k_B T)^2}],
\end{aligned}
\tag{1}
$$

where k_B is the Boltzmann constant and $T = 300\ K$ is the temperature; ω is the frequency of incident wave; τ denotes the carrier relaxation lifetime; $E_f = 0.9$ eV is the Fermi energy. And We have $\tau = \mu E_f / eV_F^2$. $V_F = 10^6$ m/s is Fermi velocity. The mobility is $\mu = 10,000\ \mathrm{cm}^2/(\mathrm{V}\cdot\mathrm{s})$, which could be realized by chemical or electrostatic doping [48,49].

The substrate is assumed to be lossless with the refractive index $n = 1.4$ and the cover layer is also semi-infinite with refractive index $n = 1$ at the beginning. The permittivity of Au is described by the Drude-Lorentz dispersion model with plasma frequency $\omega_p = 1.37 \times 10^{16}\ \mathrm{s}^{-1}$ and the damping constant $\omega_\tau = 4.05 \times 10^{13}\ \mathrm{s}^{-1}$.

4. Conclusions

In summary, a mid-infrared Fano-like resonance in a hybrid metal-graphene metamaterial has been studied. The Fano resonance arises from the coupling of the broadband resonance of the metallic nanostructure and the narrowband plasmonic resonance of the graphene nanostructure. It is sharper than plasmonic resonances in pure graphene nanostructures. The resonant strength can be effectively enhanced by increasing the layer numbers of graphene. The sensing properties of the proposed metamaterial are studied and it shows a sensitivity of about 7.93 µm/RIU. Such a sensitivity is higher than most of reported graphene plasmonic sensors in the mid-IR [28,50] while the reduced line width of the Fano resonance leads to an further increased FOM of about 158.7. The proposed concept can be employed for various modified structures. As an example, we have shown a modified hybrid graphene-metal metamaterial which is relatively easier to fabricate. The simulated transmission spectra show similar Fano resonant responses. This work may stimulate the study on Fano resonances in hybrid plasmonic structures and find applications in mid-IR sensing and others areas.

Author Contributions: J.Z. (Jianfa Zhang), X.Y., Z.Z. and S.Q. conceived the idea and supervised the study. Q.H., J.Z. (Jinglan Zou), Y.H. and J.Z. (Jianfa Zhang) conducted the numerical simulations. J.Z. (Jianfa Zhang) and Q.H. wrote the manuscript. All authors contributed to the data analysis. All authors have read and agreed to the published version of the manuscript.

Funding: This research was supported by the Science and Technology Planning Project of Hunan Province [2018JJ1033, 2017RS3039], the National Natural Science Foundation of China [11674396] and the National University of Defense Technology [ZK18-03-05].

Conflicts of Interest: The authors declare no conflict of interest.

References

1. Miroshnichenko, A.E.; Flach, S.; Kivshar, Y.S. Fano resonances in nanoscale structures. *Rev. Mod. Phys.* **2010**, *82*, 2257. [CrossRef]
2. Fedotov, V.; Rose, M.; Prosvirnin, S.; Papasimakis, N.; Zheludev, N. Sharp trapped-mode resonances in planar metamaterials with a broken structural symmetry. *Phys. Rev. Lett.* **2007**, *99*, 147401. [CrossRef] [PubMed]
3. Zhang, S.; Genov, D.A.; Wang, Y.; Liu, M.; Zhang, X. Plasmon-induced transparency in metamaterials. *Phys. Rev. Lett.* **2008**, *101*, 047401. [CrossRef] [PubMed]
4. Khardikov, V.V.; Iarko, E.O.; Prosvirnin, S.L. A giant red shift and enhancement of the light confinement in a planar array of dielectric bars. *J. Opt.* **2012**, *14*, 035103. [CrossRef]
5. Zhang, J.; MacDonald, K.F.; Zheludev, N.I. Near-infrared trapped mode magnetic resonance in an all-dielectric metamaterial. *Opt. Express* **2013**, *21*, 26721. [CrossRef]

6. Prosvirnin, S.L.; Dmitriev, V.A.; Kuleshov, Y.M.; Khardikov, V.V. Planar all-silicon metamaterial for terahertz applications. *Appl. Opt.* **2015**, *54*, 3986. [CrossRef]
7. Miroshnichenko, A.E.; Kivshar, Y.S. Fano resonances in all-dielectric oligomers. *Nano Lett.* **2012**, *12*, 6459. [CrossRef]
8. Hao, F.; Sonnefraud, Y.; Dorpe, P.V.; Maier, S.A.; Halas, N.J.; Nordlander, P. Symmetry breaking in plasmonic nanocavities: Subradiant lspr sensing and a tunable fano resonance. *Nano Lett.* **2008**, *8*, 3983. [CrossRef]
9. Hao, F.; Nordlander, P.; Sonnefraud, Y.; Dorpe, P.V.; Maier, S.A. Tunability of subradiant dipolar and fano-type plasmon resonances in metallic ring/disk cavities: Implications for nanoscale optical sensing. *ACS Nano* **2009**, *3*, 643. [CrossRef]
10. Verellen, N.; Sonnefraud, Y.; Sobhani, H.; Hao, F.; Moshchalkov, V.V.; Dorpe, P.V.; Nordlander, P.; Maier, S.A. Fano resonances in individual coherent plasmonic nanocavities. *Nano Lett.* **2009**, *9*, 1663. [CrossRef]
11. Khanikaev, A.B.; Wu, C.; Shvets, G. Fano-resonant metamaterials and their applications. *Nanophotonics* **2013**, *2*, 247. [CrossRef]
12. Wu, P.C.; Hsu, W.L.; Chen, W.T.; Huang, Y.W.; Liao, C.Y.; Liu, A.Q.; Zheludev, N.I.; Sun, G.; Tsai, D.P. Plasmon coupling in vertical split-ring resonator metamolecules. *Sci. Rep.* **2015**, *5*, 1–5. [CrossRef] [PubMed]
13. Limonov, M.F.; Rybin, M.V.; Poddubny, A.N.; Kivshar, Y.S. Fano resonances in photonics. *Nat. Photonics* **2017**, *11*, 543. [CrossRef]
14. Luk'yanchuk, B.; Zheludev, N.I.; Maier, S.A.; Halas, N.J.; Nordlander, P.; Giessen, H.; Chong, C.T. The fano resonance in plasmonic nanostructures and metamaterials. *Nat. Mater.* **2010**, *9*, 707. [CrossRef]
15. Chen, C.-Y.; Un, I.-W.; Tai, N.-H.; Yen, T.-J. Asymmetric coupling between subradiant and superradiant plasmonic resonances and its enhanced sensing performance. *Opt. Express* **2009**, *17*, 15372. [CrossRef]
16. Lahiri, B.; Khokhar, A.Z.; Richard, M.; McMeekin, S.G.; Johnson, N.P. Asymmetric split ring resonators for optical sensing of organic materials. *Opt. Express* **2009**, *17*, 1107. [CrossRef]
17. Cubukcu, E.; Zhang, S.; Park, Y.-S.; Bartal, G.; Zhang, X. Split ring resonator sensors for infrared detection of single molecular monolayers. *Appl. Phys. Lett.* **2009**, *95*, 043113. [CrossRef]
18. Kuznetsov, A.I.; Evlyukhin, A.B.; Gonçalves, M.R.; Reinhardt, C.; Koroleva, A.; Arnedillo, M.L.; Kiyan, R.; Marti, O.; Chichkov, B.N. Laser fabrication of large-scale nanoparticle arrays for sensing applications. *ACS Nano* **2011**, *5*, 4843. [CrossRef]
19. Zhang, J.; Liu, W.; Zhu, Z.; Yuan, X.; Qin, S. Strong field enhancement and light-matter interactions with all-dielectric metamaterials based on split bar resonators. *Opt. Express* **2014**, *22*, 30889. [CrossRef]
20. Wu, C.; Khanikaev, A.B.; Adato, R.; Arju, N.; Yanik, A.A.; Altug, H.; Shvets, G. Fano-resonant asymmetric metamaterials for ultrasensitive spectroscopy and identification of molecular monolayers. *Nat. Mater.* **2012**, *11*, 69. [CrossRef]
21. Grigorenko, A.; Polini, M.; Novoselov, K. Graphene plasmonics. *Nat. Photonics* **2012**, *6*, 749. [CrossRef]
22. Low, T.; Avouris, P. Graphene plasmonics for terahertz to mid-infrared applications. *ACS Nano* **2014**, *8*, 1086. [CrossRef] [PubMed]
23. Nemilentsau, A.; Low, T.; Hanson, G. Anisotropic 2d materials for tunable hyperbolic plasmonics. *Phys. Rev.* **2016**, *116*, 066804. [CrossRef] [PubMed]
24. Chen, J.; Badioli, M.; Alonso-González, P.; Thongrattanasiri, S.; Huth, F.; Osmond, J.; Spasenović, M.; Centeno, A.; Pesquera, A.; Godignon, P.; et al. Optical nano-imaging of gate-tunable graphene plasmons. *Nature* **2012**, *487*, 77. [CrossRef]
25. Fei, Z.; Rodin, A.; Andreev, G.O.; Bao, W.; McLeod, A.; Wagner, M.; Zhang, L.; Zhao, Z.; Thiemens, M.; Dominguez, G.; et al. Gate-tuning of graphene plasmons revealed by infrared nano-imaging. *Nature* **2012**, *487*, 82. [CrossRef]
26. Ju, L.; Geng, B.; Horng, J.; Girit, C.; Martin, M.; Hao, Z.; Bechtel, H.A.; Liang, X.; Zettl, A.; Shen, Y.R.; et al. Graphene plasmonics for tunable terahertz metamaterials. *Nat. Nanotechnol.* **2011**, *6*, 630. [CrossRef]
27. Yan, H.; Li, X.; Chandra, B.; Tulevski, G.; Wu, Y.; Freitag, M.; Zhu, W.; Avouris, P.; Xia, F. Tunable infrared plasmonic devices using graphene/insulator stacks. *Nat. Nanotechnol.* **2012**, *7*, 330. [CrossRef]
28. Vasić, B.; Isić, G.; Gajić, R. Localized surface plasmon resonances in graphene ribbon arrays for sensing of dielectric environment at infrared frequencies. *J. Appl. Phys.* **2013**, *113*, 013110. [CrossRef]
29. Rodrigo, D.; Limaj, O.; Janner, D.; Etezadi, D.; Abajo, F.J.G.D.; Pruneri, V.; Altug, H. Mid-infrared plasmonic biosensing with graphene. *Science* **2015**, *349*, 165. [CrossRef]

30. Wu, T.; Luo, Y.; Wei, L. Mid-infrared sensing of molecular vibrational modes with tunable graphene plasmons. *Opt. Lett.* **2017**, *42*, 2066. [CrossRef]
31. Hu, H.; Yang, X.; Guo, X.; Khaliji, K.; Biswas, S.R.; de Abajo, F.J.G.; Low, T.; Sun, Z.; Dai, Q. Gas identification with graphene plasmons. *Nat. Commun.* **2019**, *10*, 1131. [CrossRef] [PubMed]
32. Zhang, J.; Zhu, Z.; Liu, W.; Yuan, X.; Qin, S. Towards photodetection with high efficiency and tunable spectral selectivity: graphene plasmonics for light trapping and absorption engineering. *Nanoscale* **2015**, *7*, 13530. [CrossRef] [PubMed]
33. Cai, X.; Sushkov, A.B.; Jadidi, M.M.; Nyakiti, L.O.; Myers-Ward, R.L.; Gaskill, D.K.; Murphy, T.E.; Fuhrer, M.S.; Drew, H.D. Plasmon-enhanced terahertz photodetection in graphene. *Nano Lett.* **2015**, *15*, 4295. [CrossRef] [PubMed]
34. Yan, Z.; Qian, L.; Zhan, P.; Wang, Z. Generation of tunable double fano resonances by plasmon hybridization in graphene–metal metamaterial. *Appl. Phys. Express* **2018**, *11*, 072001. [CrossRef]
35. Echtermeyer, T.; Britnell, L.; Jasnos, P.; Lombardo, A.; Gorbachev, R.; Grigorenko, A.; Geim, A.; Ferrari, A.C.; Novoselov, K. Strong plasmonic enhancement of photovoltage in graphene. *Nat. Commun.* **2011**, *2*, 458. [CrossRef]
36. Xiong, F.; Zhang, J.; Zhu, Z.; Yuan, X.; Qin, S. Ultrabroadband, more than one order absorption enhancement in graphene with plasmonic light trapping. *Sci. Rep.* **2015**, *5*, 16998. [CrossRef]
37. Liu, P.Q.; Luxmoore, I.J.; Mikhailov, S.A.; Savostianova, N.A.; Valmorra, F.; Faist, J.; Nash, G.R. Highly tunable hybrid metamaterials employing split-ring resonators strongly coupled to graphene surface plasmons. *Nat. Commun.* **2015**, *6*, 8969. [CrossRef]
38. Yan, X.; Wang, T.; Xiao, S.; Liu, T.; Hou, H.; Cheng, L.; Jiang, X. Dynamically controllable plasmon induced transparency based on hybrid metal-graphene metamaterials. *Sci. Rep.* **2017**, *7*, 13917. [CrossRef]
39. Xiao, S.; Wang, T.; Liu, T.; Yan, X.; Li, Z.; Xu, C. Active modulation of electromagnetically induced transparency analogue in terahertz hybrid metal-graphene metamaterials. *Carbon* **2018**, *126*, 271. [CrossRef]
40. Hong, Q.; Luo, J.; Wen, C.; Zhang, J.; Zhu, Z.; Qin, S.; Yuan, X. Hybrid metal-graphene plasmonic sensor for multi-spectral sensing in both near-and mid-infrared ranges. *Opt. Express* **2019**, *27*, 35914. [CrossRef]
41. Chen, Z.-x.; Chen, J.-h.; Wu, Z.-j.; Hu, W.; Zhang, X.-j.; Lu, Y.-q. Tunable fano resonance in hybrid graphene-metal gratings. *Appl. Phys. Lett.* **2014**, *104*, 161114. [CrossRef]
42. Pan, M.; Liang, Z.; Wang, Y.; Chen, Y. Tunable angle-independent refractive index sensor based on fano resonance in integrated metal and graphene nanoribbons. *Sci. Rep.* **2016**, *6*, 29984. [CrossRef] [PubMed]
43. Rodrigo, D.; Tittl, A.; Limaj, O.; Abajo, F.J.G.D.; Pruneri, V.; Altug, H. Complete optical absorption in periodically patterned graphene. *Light Sci. Appl.* **2017**, *6*, e16277. [CrossRef] [PubMed]
44. Thongrattanasiri, S.; Koppens, F.H.; de Abajo, F.J.G. Broad electrical tuning of graphene-loaded plasmonic antennas. *Phys. Rev. Lett.* **2012**, *108*, 047401. [CrossRef]
45. Yao, Y.; Kats, M.A.; Genevet, P.; Yu, N.; Song, Y.; Kong, J.; Capasso, F. Optical far-infrared properties of a graphene monolayer and multilayer. *Nano Lett.* **2013**, *13*, 1257. [CrossRef]
46. Falkovsky, L.; Pershoguba, S. Space-time dispersion of graphene conductivity. *Phys. Rev. B* **2007**, *76*, 153410. [CrossRef]
47. Falkovsky, L.; Varlamov, A. Electric field effect in atomically thin carbon films. *Eur. Phys. J. B* **2007**, *56*, 281. [CrossRef]
48. Novoselov, K.S.; Geim, A.K.; Morozov, S.; Jiang, D.; Zhang, Y.; Dubonos, S.; Grigorieva, I.; Firsov, A. Gated tunability and hybridization of localized plasmons in nanostructured graphene. *Science* **2004**, *306*, 666. [CrossRef]
49. Fang, Z.; Thongrattanasiri, S.; Schlather, A.; Liu, Z.; Ma, L.; Wang, Y.; Ajayan, P.M.; Nordlander, P.; Halas, N.J.; de Abajo, F.J.G. Double-layer graphene for enhanced tunable infrared plasmonics. *ACS Nano* **2013**, *7*, 2388. [CrossRef]
50. Wenger, T.; Viola, G.; Kinaret, J.; Fogelström, M.; Tassin, P. High-sensitivity plasmonic refractive index sensing using graphene. *2D Mater.* **2017**, *4*, 025103. [CrossRef]

Article

Morphology Evolution of Nanoscale-Thick Au/Pd Bimetallic Films on Silicon Carbide Substrate

Francesco Ruffino *, Maria Censabella, Giovanni Piccitto and Maria Grazia Grimaldi

Dipartimento di Fisica e Astronomia "Ettore Majorana", Università di Catania and MATIS CNR-IMM, via S. Sofia 64, 95123 Catania, Italy; maria.censabella@ct.infn.it (M.C.); giovanni.piccitto@ct.infn.it (G.P.); mariagrazia.grimaldi@ct.infn.it (M.G.G.)

* Correspondence: francesco.ruffino@ct.infn.it

Received: 28 March 2020; Accepted: 13 April 2020; Published: 14 April 2020

Abstract: Bimetallic Au/Pd nanoscale-thick films were sputter-deposited at room temperature on a silicon carbide (SiC) surface, and the surface-morphology evolution of the films versus thickness was studied with scanning electron microscopy. This study allowed to elucidate the Au/Pd growth mechanism by identifying characteristic growth regimes, and to quantify the characteristic parameters of the growth process. In particular, we observed that the Au/Pd film initially grew as three-dimensional clusters; then, increasing Au/Pd film thickness, film morphology evolved from isolated clusters to partially coalesced wormlike structures, followed by percolation morphology, and, finally, into a continuous rough film. The application of the interrupted coalescence model allowed us to evaluate a critical mean cluster diameter for partial coalescence, and the application of Vincent's model allowed us to quantify the critical Au/Pd coverage for percolation transition.

Keywords: Au/Pd; SiC; nanomorphology; coalescence; percolation; scanning electron microscopy

---

1. Introduction

Silicon carbide (SiC) is a semiconductor, and ceramic material that has interesting physical and chemical properties that are useful for various applications in different technological areas as a structural material in electronics and optoelectronics [1–3]. In these applications, SiC's characteristic properties, such as a high melting temperature, high thermal conductivity, high Young modulus, wide energy band gap, and chemical stability are fully exploited [1–3]. In particular, metal/SiC Schottky diodes are used in the fabrication of different devices, ranging from high-temperature and -power electronics to very sensitive high-temperature hydrogen and hydrocarbon gas detectors and biosensors (due to SiC's biocompatibility properties) [1–15]. So, in the last few decades, to meet specific technological requirements, various methodologies were exploited for the fabrication of metal/SiC diodes. In particular, all te elemental metals and, often, combinations of metals were used to produce SiC-based Ohmic and Schottky contacts with resulting characteristics of barrier height, stability, etc. being widely investigated. In particular, Pd/SiC [1–7,9,11,13,16] and Au/SiC [1–6,16–22] Schottky contacts with metals in the form of nanoscale-thick deposited films or complex-shape deposited nanostructures attracted much interest for various technological applications ranging from electronics and optoelectronics to sensing and catalysis. In fact, in general, nanoscale-thick metal films and metal nanostructures on functional surfaces are the basis of innovative versatile and high-performance devices [23–30], whose properties are strongly dependent on the morphology, structure, shape, size, and metal–substrate interactions of metal films and nanostructures. Nanoscale metals, in fact, exhibit valuable properties that change by size reduction arising from the electron-confinement effect, and variations of electronic structure and surface effects [23–26]. These properties are size-, shape-, and composition-dependent, and are exploited in plasmonic, sensing, electrical, and catalytic applications [23–26]. In addition, regarding nanoscale metals supported on substrates, a specific metal–substrate interaction often leads to an overall composite whose properties arise from the synergistic combination of the properties of individual components, resulting in new functional characteristics. In this regard, nanoscale metals/SiC combinations have potential to reach a high level of efficiency, overcoming the major technical problem related to the low operating temperature of standard metals/Si devices [1–22]. In particular, nanoscale-thick Pd and Au films or Pd and Au nanostructures fabricated on a SiC surface showed peculiar and interesting properties that are useful, for example, in gas sensing [7,31], biosensing [21,22], and catalysts [32]. Besides pure metals, bimetallic nanoscale metals have shown to have improved characteristics with respect to the characteristics of their individual components [25,33]. In particular, for example, Pd is an excellent catalyst for many reactions with improved effects in nanoscale form [34–37]. Furthermore, the catalytic activity of nanoscale Pd is often enhanced by the addition of Au [37,38]. In this regard, bimetallic Au/Pd nanoparticles on a SiC surface have recently showed an efficient photocatalytic effect in the hydrogenation of nitroarenes, with a key role played by the catalytic properties of bimetallic Au/Pd nanoparticles and of the specific energy-band diagram of SiC [39].

In general, towards real device applications, the critical issue is the controlled direct fabrication of nanoscale metals with the desired structure and morphology on the support substrate. Exploiting physical-vapor deposition approaches for the formation of nanoscale metals on substrates, this can be achieved by controlling the deposition-process parameters once the basic microscopic kinetics and thermodynamics mechanisms, governing the metal growth, are known and related to the growing film's nanoscale morphology. This control ensures the desired nanomorphology control for specific applications.

On the basis of these considerations, we report on the study of the morphological characteristics of bimetallic Au/Pd nanoscale-thick films sputter-deposited on a SiC substrate. In particular, using the sputter-deposition technique, we deposited bimetallic Au/Pd films on the SiC surface increasing overall film thickness from ~1 to ~10 nm, and we used the scanning-electron-microscopy (SEM) technique to study the surface-morphology evolution of the Au/Pd film versus film thickness.

The common methods of metal deposition used in SiC device fabrication are sputtering, evaporation, chemical-vapor deposition, and atomic-layer deposition, in order of applicability [40]. However, sputtering is most commonly used today, since it is a fast and economical method that can meet industrial requirements. In addition, metal adhesion is good, and compound targets are possible, as in the case of the Au/Pd target. Evaporation is also common, since it allows higher deposition rates and can be performed in an ultraclean vacuum system. Higher demands are usually placed on the vacuum system, and adhesion can be a problem [40]. Metals can also be deposited by chemical-vapor and atomic-layer deposition, but they are not as common as the other two. An advantage with the chemical-vapor and atomic-layer deposition methods is that epitaxial growth is easier to promote, because of the thermal energy present during the deposition; however, reports on SiC metallization using these techniques are rare [40,41] (instead, for example, atomic-layer deposition is widely used to deposit gate dielectrics on SiC to obtain high-quality interfaces [42]). For these reasons, here we focus on the study of the morphology evolution of the Au/Pd films deposited on SiC substrates by sputtering techniques: results could have a direct connection to industrial technologically important applications.

First, this study allowed us to identify different growth regimes for the Au/Pd film increasing the film thickness: in the initial growth stage (nucleation regime), the nanostructured Au/Pd film was formed by nanoscale dropletlike clusters that evolved into wormlike islands by an interrupted coalescence process, increasing the amount of the deposited metals; further increasing film thickness, a percolation stage was reached for which a discontinuous holed Au/Pd film was obtained; finally, further increasing film thickness, the film became a compact and continuous rough layer. SEM images were quantitatively analyzed to extract the evolution of the mean sizes and surface density of the Au/Pd droplets and islands versus film thickness, and to extract the surface coverage of the Au/Pd film versus film thickness. The plot of the Au/Pd droplets and wormlike-island sizes was analyzed using the interrupted coalescence model to quantify the critical Au/Pd droplet size; after that, the droplets' nucleation stage ended, and interrupted coalescence growth leading to the formation of the wormlike structures started. The plots of the droplet and island surface density and surface coverage versus film thickness were analyzed by Vincent's model to quantify the critical Au/Pd surface coverage, after which the percolation stage started.

These data established a general working framework quantitatively connecting the Au/Pd nanoscale morphology to film thickness, giving the possibility to choose specific deposition conditions to obtain the desired nanoscale morphology of the Au/Pd film for specific applications. These results could be generally useful to improve metallization schemes for SiC, and find applications in some recent technological development, such as in analytical fields related to nanostructure laser desorption/ionization (NALDI) [43,44], and in the epitaxial growth of graphene on SiC with metal intercalation [45,46]. In fact, surface-assisted laser desorption/ionization time-of-flight mass spectrometry using nanoparticles and nanostructured surfaces take the advantages of minimal fragmentation of analytes resulting in extremely high sensitivity. In this regard, nanostructured metal films, as in the case of Au/Pd on SiC, could be very useful to further improve the technique. On the other hand, the application of graphene in electronic devices requires large-scale epitaxial growth. The presence of the substrate, however, worsens charge-carrier mobility. Several works demonstrated the possibility to decouple the partially sp^3-hybridized first graphitic layer formed on the Si-terminated face of SiC from the substrate by metal (i.e., gold) intercalation, leading to completely sp^2-hybridized graphene layer with improved electronic properties. A buffer layer on the SiC surface during graphene growth could meet this approach.

2. Materials, Experiment Analysis, and Methods

The used substrates were 1 × 1 cm n-type 6H–SiC pieces (Si-terminated, doping concentration $N_D \approx 5.1 \times 10^{17}$ cm^{-3}). The Au/Pd films were deposited on the SiC piece surface by sputter deposition. For these depositions, an $Au_{60}Pd_{40}$ (atomic %, nominal composition) target (99.999% purity) was used with a circular shape and diameter of 6 cm. The SiC pieces were located in front of the target at a distance of 4 cm. During the depositions, Ar (0.02 mbar) was used as the working gas. The emission current was fixed to 10 mA (so, the applied power was consequently fixed), and the deposition time was varied to change the effective thickness of the deposited Au/Pd film, the deposition rate being ~1.2 nm/min. However, after the deposition processes, the resulting effective thicknesses of the films were checked by ex situ Rutherford backscattering analysis (RBS) performed on the Au/Pd films deposited on reference Si substrates (to easily interpret the resulting RBS spectra) on which the films were deposited in the same running processes of the SiC substrates. These analyses were performed by using 2 MeV $^4He^+$ backscattered ions with a scattering angle of 165°. RUMP simulations [47] of the RBS spectra first indicated the effective composition of the deposited films as $Au_{64}Pd_{36}$ (atomic %). Then, the RUMP simulations of the RBS spectra indicated the following effective thicknesses for the deposited Au/Pd films: h = 1.2, 2.1, 2.8, 3.5, 4.5, 5.2, 6.9, 8.8, and 10.5 nm with a measurement error of 5%. In particular, RBS analyses allowed to separately evaluate the amount of deposited Au and Pd in atoms/cm^2 that were converted into Au and Pd thickness by dividing the corresponding measured amount for metal atomic density (5.9×10^{22} atoms/cm^3 in the case of Au; 6.8×10^{22} atoms/cm^3 in the case of Pd) weighted for the corresponding metal atomic percent in the target (0.64 for Au and 0.36 for Pd), and finally summed to give the resulting overall effective thickness of the Au/Pd film. As an example, Figure 1 reports the experimental RBS spectrum (black full line) concerning the Si/Au/Pd where the Au/Pd film resulted in 4.5 nm thickness. The red full line indicates the corresponding simulated spectrum.

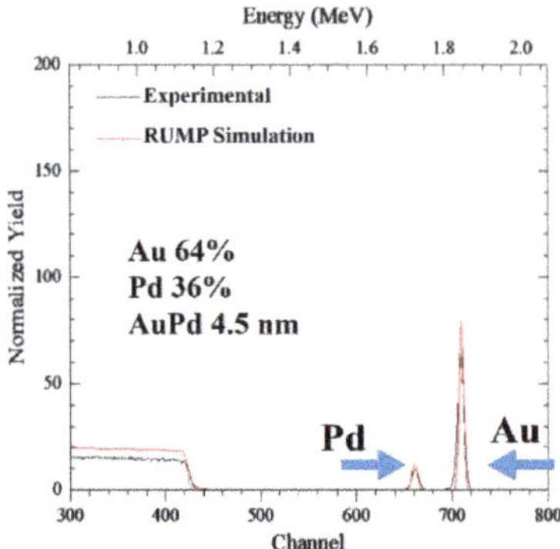

Figure 1. Representative Rutherford backscattering analysis (RBS) spectrum: black line, experimental spectrum; red line, RUMP simulation. Spectrum refers to a Au/Pd film deposited on reference Si substrate. Peaks corresponding to corresponding to presence of Au and Pd are indicated. Simulation of experimental spectrum indicates an atomic composition of 64% Au and 36% Pd film. Correspondingly, an effective thickness for the Au/Pd film of 4.5 nm was evaluated in this case.

Morphological studies were mainly performed with scanning electron microscopy (SEM) employing a Gemini Field Emission Carl Zeiss SUPRA 25 microscope (Carl Zeiss, Oberkochen, Germany). In SEM measurements, a 5 kV accelerated electron beam was used, and a working distance of 3 mm was set. Gatan Digital Micrograph software (Version 3.4.1, Gatan Inc., Ametek, Berwin, PA, USA) was employed to analyze the acquired SEM images; in particular, the brightness of the image was manipulated so to obtain the brighter regions (intensity value set to 1) corresponding to Au/Pd surface areas, and the darker regions (intensity value 0) corresponding to the SiC surface areas. In this way, when identifying particles of a circular cross-section in the SEM image, diameter (width = length = diameter) R of the circle was measured for at least 200 particles per sample to construct sizes distributions from which mean diameter <R> and the corresponding error (evaluated as standard deviation) were derived. On the other hand, when identifying islands with an elongated cross-section, the smaller ellipse inscribing the particle for each structure was considered from which width R and length D of the island (ellipse) were measured; this was repeated for at least 200 islands per sample to construct size distributions from which mean width <R>, mean length <D>, and corresponding errors (evaluated as standard deviations) were derived. Surface density N (structures/cm^2) of the Au/Pd particles or islands was evaluated by direct counting and averaging on at least 5 SEM images per sample (and the corresponding error arising from the averaging procedure). Surface coverage P of the Au/Pd film was evaluated as the ratio between the occupied area in an SEM image by the film (bright area) and the total area of the image (bright + dark areas). This procedure was repeated for at least 5 SEM images per sample, and the values of the surface coverage and the corresponding errors were evaluated by the averaging procedure.

In addition, atomic-force-microscopy (AFM) analyses were carried out by employing a Bruker-Innova microscope (Billerica, MA, USA). The measurements were performed in high-amplitude mode. MSNL-10 Si tips were used (from Bruker) having curvature radius of ~2 nm. AFM images were analyzed by using SPMLABANALYSES V7.00 software (Version 7, Billerica, MA, USA). AFM images could be similarly analyzed to the SEM images (using, also, line-section profiles of the surface structures) to extract quantitative information on the size of the surface Au/Pd nanostructures, their surface density, and the fraction of the surface area covered by Au/Pd.

In the best setup conditions, the Gemini Field Emission Carl Zeiss SUPRA 25 microscope was in ~1.7 nm lateral resolution [48]. However, in the experiment setup, conditions for SEM measurements in lateral SEM resolution were comparable to those of the AFM measurements (~2 nm, the radius of curvature for the used Si tips) since independent analysis of SEM and AFM images resulted in very similar results for the nanoparticle sizes and surface densities, and for the fraction of surface area covered by Au/Pd. Results obtained by SEM analysis checked by independent AFM measurements are reported below.

3. Results and Discussion

Figure 2 shows the AFM and SEM images of the starting SiC bare surface: (a) 2 × 2 μm AFM image (three-dimensional reconstruction) of SiC surface, (b,c) SEM images of SiC surface (increasing magnification from (a) to (b)). No particular surface features were recognizable either on a large scale (low-magnification image in (a)) or on a small scale (high magnification in (b)). The SiC surface was clean and flat. In this regard, AFM analysis allowed to evaluate that the root-mean square (RMS, i.e., standard deviation of surface heights) equaled to 0.2 nm, confirming the flatness of the starting SiC surface.

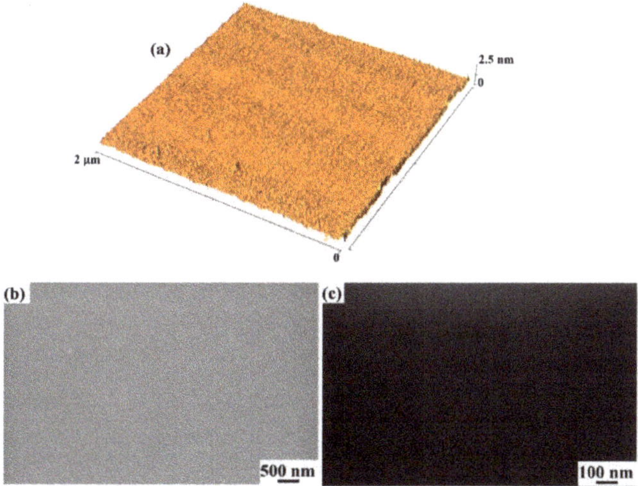

Figure 2. (**a**) A 2 μ × 2 μm AFM image (three-dimensional reconstruction) of starting bare silicon carbide (SiC) surface. (**b**,**c**) SEM images of starting bare SiC surface with magnification increasing from (**b**) to (**c**).

Figure 3 reports the representative SEM images of the surface of the Au/Pd film deposited on the SiC surface for different film thicknesses h: (a–c) h = 1.2 nm and magnification increasing from (a) to (c); (d–f) h = 3.5 nm and magnification increasing from (d) to (f); (g–i) h = 6.9 nm and magnification increasing from (g) to (i); (j–l) h = 10.5 nm and magnification increasing from (j) to (l).

Concerning the deposited Au/Pd films with thickness $h < 3.5$ nm, the film is formed by small clusters (nuclei) with a circular cross-section, as can be recognized from the SEM images in Figure 3a–c corresponding to the film with thickness h = 1.2 nm. This is the very first stage of metal growth denoted as the nucleation stage. Metal atoms or small clusters deposited on the substrate surface from the vapor phase are characterized by mobility that leads them to probe a certain surface area and meet other deposited atoms, clusters, or surface defects to start the nucleation process of small three-dimensional islands that grow in size as metal deposition continues [49–51]. Both Au and Pd have a strong nonwetting nature on the SiC surface [16,18,52], so starting from the bimetallic Au/Pd target used for the depositions, the Au/Pd film grows in this nucleation regime, following Volmer–Weber growth [51] mode resulting in the formation of a three-dimensional cluster with dropletlike shapes (spherical or hemispherical), with their sizes represented by their diameters R. Increasing Au/Pd film thickness, and in particular for $3.5 \leq h < 5.2$ nm, we observed a transition of the cluster size from the circular cross-sectional shape to an elongated cross shape, as evident from the SEM images in Figure 3d–f corresponding to the film with h = 5.2 nm. In this thickness range, the metal clusters became two-dimensional yet separated islands due to a (partial) coalescence process of the growing metal clusters leading to the formation of the wormlike structures. This shape transition is, roughly, represented in Figure 4, evidencing that, for the elongated metal islands, the two characteristic planar sizes were width R and length D. Such growth evolution is typical for metal films on nonwetting substrates [16,53–59], described by the interrupted coalescence model (ICM) first proposed by Yu et al. [53]. Starting from the dropletlike metal clusters obtained in the nucleation stage, further increasing the amount of deposited metal atoms, the growing clusters can touch each other, giving origin to a coalescence process (two or more clusters merge to form a unique cluster with volume equal to the sum of the volumes of the contacting clusters, but with a lower surface area than the sum of the surface areas of the contacting clusters) to minimize the total surface area of the system. Therefore, at this stage, the thermodynamic change drives system transformation towards the minimum of surface energy. However, in this coalescence process, the substrate can act by a wiping action leading to a partial coalescence process in which part of the substrate, which was initially covered by metal, is then wiped clean. As a result, the metal film is formed by elongated islands resulting from full coalescence in one direction and partial coalescence in another direction. The islands are separated by gaps between them. The ICM model identified critical diameter R_c for the dropletlike growing metal clusters separating the nucleation and growth stages to partial (interrupted) coalescence, being R_c-dependent on the chemical nature of the metal and substrate, and on temperature.

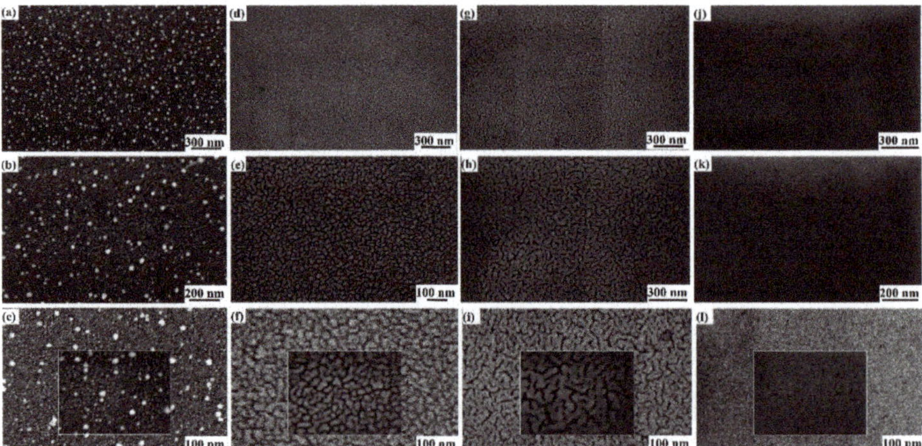

Figure 3. Examples of SEM images of surface of Au/Pd film deposited on SiC surface for different film thicknesses h: (**a–c**) h = 1.2 nm and magnification increasing from (**a**) to (**c**); (**d–f**) h = 3.5 nm and magnification increasing from (**d**) to (**f**); (**g–i**) h = 6.9 nm and magnification increasing from (**g**) to (**i**); (**j–l**) h = 10.5 nm and magnification increasing from (**j**) to (**l**). In (**c**), (**f**), (**i**), and (**l**), marker refers to the image in the smaller rectangle where, at parity of the electron-beam scanning rate, a smaller surface area was imaged to reach a better lateral resolution in higher-magnification analysis.

Further increasing the amount of deposited metals with respect to the partial-coalescence stage, a percolation regime occurs, in our case, for Au/Pd film thickness of $5.2 < h \leq 8.8$ nm. Percolative morphology for the Au/Pd film can be observed in the SEM images in Figure 3g–i corresponding to the film with h = 6.9 nm. In this growth regime, the increased amount of deposited metal led to the continuous growth of the two-dimensional coalesced islands, which grew much longer than larger until connecting with each other to form a quasicontinuous (holed) network across the surface. Finally, further increasing the amount of deposited metal, a compact, rough, continuous film was obtained by a filling process of the holes that were present in the previously obtained percolative film. In our case, this occurred for 8.8 < h ≤ 10.5 nm. As an example, the continuous, rough Au/Pd film can be recognized in the SEM images in Figure 3j–l corresponding to the film with h = 10.5 nm.

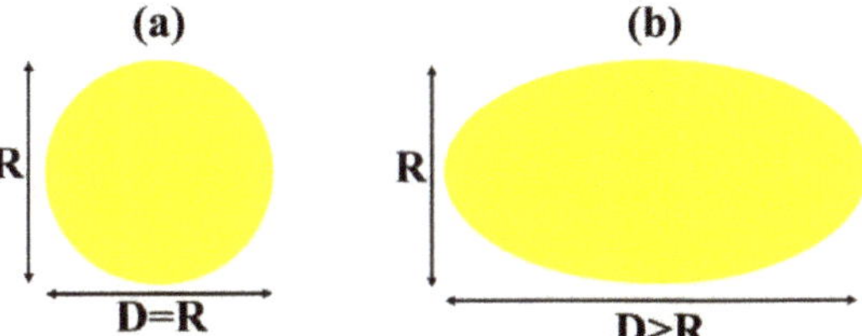

Figure 4. Planar section of an Au/Pd cluster in (**a**) first growth stages (a circle of diameter D) and (**b**) later growth stages (elongated shape represented as an ellipse of characteristic sizes R and D > R).

We can characterize the partial-coalescence and percolation stages by the quantitative evaluation and analysis of some parameters. First, we used the SEM images (see Section 2) to derive the values for the mean sizes of the Au/Pd surface structures versus film thickness. In the nucleation stage, we evaluated diameter R of the circular cross-section of the dropletlike Au/Pd particles to construct diameter distributions from which mean diameter <R> (and the corresponding standard deviation) was extracted. As an example, Figure 5a,b reports, respectively, the distribution of width R and length D for the dropletlike particles observed in the sample with an Au/Pd film thickness of h = 2.8 nm; <R> = <D> ≈ 19.5 nm, confirming the circular cross-section of the Au/Pd structures characterized by unique mean diameter <R>. In addition, in the partial-coalescence stage, we evaluated width R and length D of the partially coalesced wormlike Au/Pd structures to construct the distributions for R and D, from which the mean values <R> and <D> (and the corresponding standard deviations) were extracted. An example is shown in Figure 5c,d, where the distributions for R and D corresponding to the Au/Pd structures were observed in the Au/Pd film with thickness of h = 5.2 nm. In this case, <R> = 42.5 nm and <D> = 80.3 nm, confirming that, in the partial-coalescence stage, (<D>) > (<R>), a condition characteristic of elongated Au/Pd structures. Overall, Figure 6 reports (on a log–log scale) the experimentally derived values (dots) of mean island width <R> and length <D> as a function of film thickness h (for 1.2 ≤ h ≤ 5.2 nm). For h = 1.2, 2.1, and 2.8 nm, condition <R> ≈ <D> was realized; for h = 3.5, 4.5, and 5.2 nm, condition (<D>) > (<R>). In the log–log scale of the plot in Figure 6, data <R> versus h and <D> versus h were fitted by linear functions (black full line for <R> versus h with 1.2 ≤ h ≤ 5.2 nm, red full line for <D> versus h with 3.5 ≤ h ≤ 5.2 nm). The intersection of the two full lines allowed to evaluate critical radius R_c defined by the ICM model from which the partial-coalescence process of the Au/Pd clusters began.

So, we obtained R_c = 23.3 nm corresponding to the critical thickness for the Au/Pd film h_c = 3.2 nm. Therefore, at critical thickness 3.2 nm of the deposited Au/Pd film (corresponding to a mean critical Au/Pd cluster diameter of 23.3 nm), the nucleation and growth stages of the dropletlike Au/Pd clusters ended, and the partial-coalescence stage started. Continuing the metal deposition, a later stage of the Au/Pd film growth occurred: the elongated islands grew larger and longer, and when the peripheries of the neighboring islands met and touched, the onset for percolation occurred. To characterize the percolation-growth stage, we used Vincent's model [59–61]: in this model, the transition from isolated islands to percolation occurs at critical film-percolation coverage P_c. In the framework of Vincent's model, film-percolation coverage P is related to metal-island surface density N by (in the hypothesis of noninstantaneous coalescence [61])

$$P = P_0(N/N_0) + P_c[1 - (N/N_0)], \tag{1}$$

where P_0, initial metal coverage; N_0, initial metal-island surface density ($N = N_0$ when $P = P_0$); and P_c, critical film coverage for which percolation occurs. The application of Equation (1) requires to know N_0. In this regard, in the framework of Vincent's model, the metal-island surface density was dependent on film thickness by [59]

$$\ln(N/N_0) = -Ah^{2/3}, \quad (2)$$

where A is a constant determined by the shape of the islands.

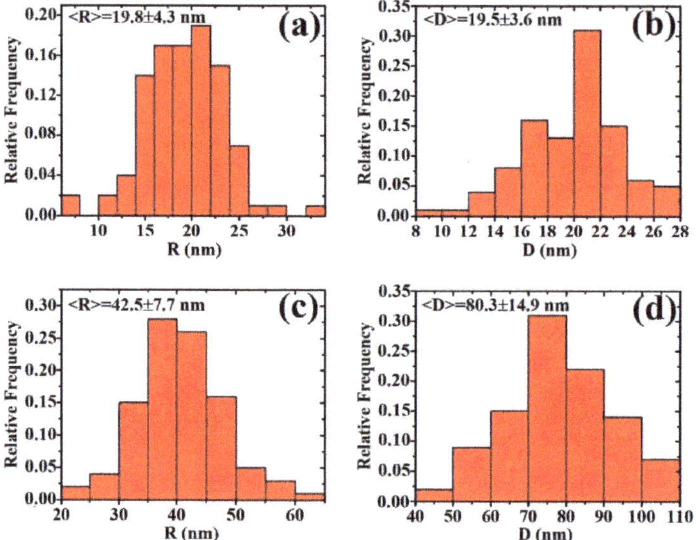

Figure 5. Representative size distributions for Au/Pd clusters: (**a**,**b**) distributions for width R and length D, respectively, of Au/Pd clusters in deposited film of h = 2.8 nm thickness; (**c**,**d**) distributions for width R and length D, respectively, of Au/Pd clusters in deposited film of h = 5.2 nm thickness. In each case, mean values of width <R> and length <D>, as extracted by the distribution, are reported.

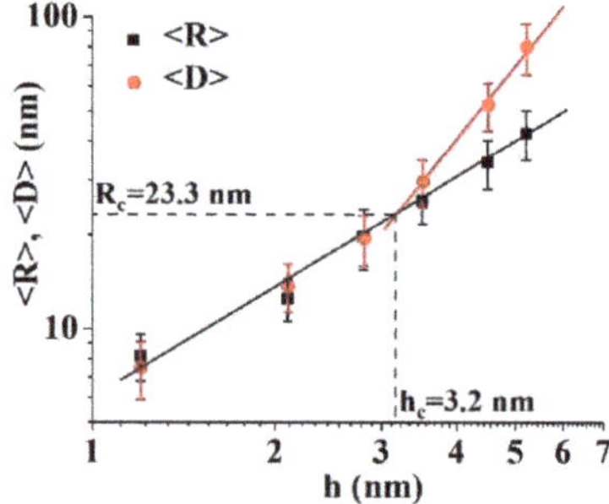

Figure 6. Evolution plot of mean width <R> and mean length <D> of Au/Pd islands versus thickness h of the deposited Au/Pd film, for 1.2 ≤ h ≤ 5.2 nm. Full lines arose by fitting procedures of experiment data.

On the basis of these theoretical results, after evaluating the mean values of surface density N for the Au/Pd clusters and islands for 1.2 ≤ h ≤ 5.2 nm, Figure 7 reports, on a semilog scale, the values of N (dots) versus $h^{2/3}$, so that, from the linear fit of the experiment data, value $N_0 = 1.9 \times 10^{12}$ cm^{-2} (considered as fitting parameter) was obtained for the initial surface density of the Au/Pd clusters.

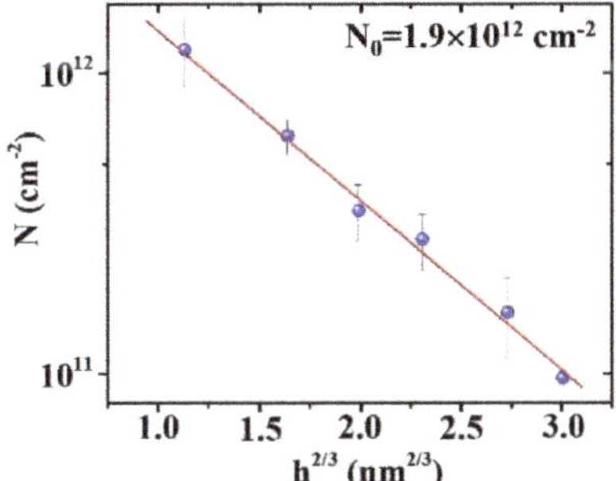

Figure 7. Plot of mean number of Au/Pd islands (LogN) versus $h^{2/3}$ with h being the thickness of the deposited Au/Pd film (with $1.2 \leq h \leq 5.2$ nm). Full line arose by the fitting procedure of the experiment data.

Once N_0 was evaluated, we reported the experiment-derived values of Au/Pd surface coverage P (evaluated for each film thickness h) versus N/N_0 (with N the Au/Pd island surface density evaluated for each corresponding film thickness h); see the dots in the plot in Figure 8. The linear fit of the experiment data (full line in Figure 8) allowed to evaluate the value of the initial Au/Pd surface coverage as fitting parameters, $P_0 = 27\%$, and, in particular, the critical value for Au/Pd surface coverage at which the percolation stage for the Au/Pd film occurred, $P_c = 61\%$.

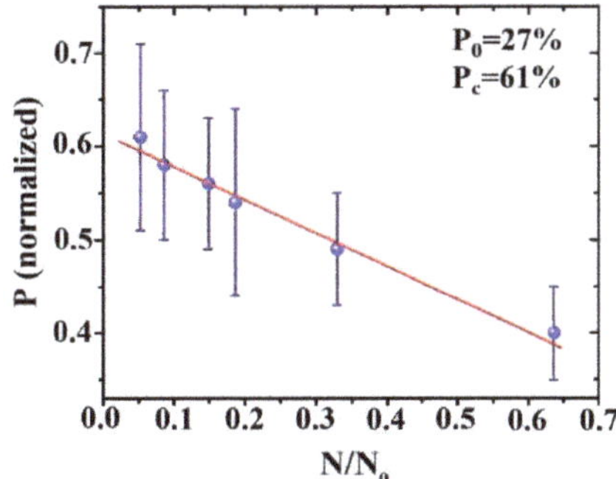

Figure 8. Plot of fraction of surface area covered by Au/Pd film P versus number N of Au/Pd islands normalized to N_0 (starting island density).

4. Conclusions

We studied the growth morphology of nanoscale-thick (1–10 nm) bimetallic Au/Pd film that was sputter-deposited, at room temperature, on 6H–SiC, versus film thickness h. Increasing film thickness, we observed the evolution of the Au/Pd film as formed in the initial stage of growth ($h < 3.5$ nm, nucleation regime) by three-dimensional isolated particles with circular cross-section to a film as formed by partially coalesced wormlike islands ($3.5 \leq h < 5.2$ nm, partial-coalescence stage), to a percolative film ($5.2 < h \leq 8.8$ nm, percolation stage), and finally to a continuous, compact, rough film ($h = 10.5$ nm). The interrupted-coalescence model was applied to describe this growth sequence, allowing to estimate critical mean Au/Pd particle diameter $R_c = 23.3$ nm (corresponding to critical film thickness $h_c = 3.2$ nm) to start the partial-coalescence regime. Furthermore, the evolution of the Au/Pd particle surface density and coverage P was analyzed by using Vincent's model to analyze, in particular, the percolation process at the later stage of growth, allowing to evaluate percolation threshold $P_c = 61\%$.

These results set a general framework that can connect Au/Pd nanoscale morphology to film thickness, giving the possibility to controllably tune film morphology for specific miniaturized-device applications.

Author Contributions: Conceptualization, F.R., G.P., and M.G.G.; methodology, F.R., G.P., and M.G.G.; formal analysis, F.R. and M.C.; investigation, F.R. and M.C.; data curation, F.R. and M.C.; writing—original-draft preparation, F.R.; writing—review and editing, F.R., M.C., G.P., and M.G.G. All authors have read and agreed to the published version of the manuscript.

Funding: This work was supported by project "Materiali innovativi e nano strutturati per microelettronica, energia e sensoristica"—Linea di intervento 2 (Univ. Catania, DFA), and by project "Contratto di Sviluppo "STMicroelectronics" (CDS_000448)".

Conflicts of Interest: The authors declare no conflict of interest.

References

1. Saddow, S.E.; Agarwal, A. *Advances in Silicon Carbide Processing and Applications*; Artech House Inc.: Norwood, MA, USA, 2004.
2. Kimoto, T.; Cooper, J.A. *Fundamentals of Silicon Carbide Technology-Growth, Characterization, Devices, and Applications*; Wiley: Singapore, 2014.
3. Shur, M.; Rumyantsev, S.; Levinshtein, M. *SiC Materials and Devices, vol. 1 and vol. 2*; World Scientific Publishing: Singapore, 2007.
4. Porter, L.M.; Davis, R.F. A critical review of ohmic and rectifying contacts for silicon carbide. *Mater. Sci. Eng. B* **1995**, *34*, 83–105. [CrossRef]
5. Zekentes, K.; Vasilevskiy, K. *Advancing Silicon Carbide Electronics Technology I-Metal Contacts to Silicon Carbide: Physics, Technology, Applications*; Materials Research Forum LLC: Millersville, PA, USA, 2018.
6. Friedrichs, P.; Kimoto, T.; Ley, L.; Pensl, G. *Silicon Carbide-Power Devices and Sensors (Vol. 2)*; Wiley: Weinheim, Germany, 2010.
7. Kim, C.K.; Lee, J.H.; Choi, S.M.; Noh, I.H.; Kim, H.R.; Cho, N.I.; Hong, C.; Jang, G.E. Pd- and Pt-SiC Schottky diodes for detection of H_2 and CH_4 at high temperature. *Sens. Actuators B* **2001**, *77*, 455–462. [CrossRef]
8. Khan, S.A.; de Vasconcelos, E.A.; Uchida, H.; Katsube, T. Gas response and modelling of NO-sensitive thin-Pt SiC Schottky diodes. *Sens. Actuators B* **2003**, *92*, 181–185. [CrossRef]
9. Olivier, E.J.; Neethling, J.H. Palladium transport in SiC. *Nucl. Eng. Des.* **2012**, *244*, 25–33. [CrossRef]
10. Yun, S.; Wang, L.; Zhao, C.; Wang, Y.; Ma, T. A new type of loss-cost counter electrode catalyst based on platinum nanoparticles loaded onto silicon carbide (Pt/SiC) for dye-sensitized solar cells. *Phys. Chem. Chem. Phys.* **2013**, *15*, 4286–4290. [CrossRef]
11. Njoroge, E.G.; Theron, C.C.; Skuratov, V.A.; Wamwangi, D.; Hlatshwayo, T.T.; Comrie, C.M. Malherbe, J.B.; Interface reactions between Pd thin films and SiC by thermal annealing and SHI irradiation. *Nucl. Instr. Meth. Phys. Res. B* **2016**, *371*, 263–267. [CrossRef]
12. Şennik, E.; Ürdem, Ş.; Erkovan, M.; Kilinç, N. Sputtered platinum thin films for resistive hydrogen sensor application. *Mater. Lett.* **2016**, *177*, 104–107. [CrossRef]
13. Thabethe, T.T.; Nstoane, T.; Biira, S.; Njoroge, E.G.; Hlatshwayo, T.T.; Skuratov, V.A.; Malherbe, J.B. Irradiation effects of swift heavy ions on palladium films deposited on 6H-SiC substrate. *Nucl. Instr. Meth. Phys. Res.* **2019**, *442*, 19–23. [CrossRef]
14. Wright, N.G.; Horsfall, A.B. SiC sensors: A review. *J. Phys. D Appl. Phys.* **2007**, *40*, 6345–6354. [CrossRef]
15. Casals, O.; Barcones, B.; Romano-Rodríguez, A.; Serre, C.; Pérez-Rodríguez, A.; Morante, J.R.; Godignon, P.; Montserrat, J.; Millán, J. Characterisation and stabilisation of Pt/TaSiX/SiO_2/SiC gas sensor. *Sens. Actuators B Chem.* **2005**, *109*, 119–127. [CrossRef]
16. Ruffino, F.; Grimaldi, M.G. Island-to-percolation transition during the room-temprature growth of sputtered nanoscale Pd films on hexagonal SiC. *J. Appl. Phys.* **2010**, *107*, 074301. [CrossRef]
17. Ruffino, F.; Grimaldi, M.G.; Giannazzo, F.; Roccaforte, F.; Raineri, V. Size-dependent Schottky barrier height in self-assembled gold nanoparticles. *Appl. Phys. Lett.* **2006**, *89*, 243113. [CrossRef]
18. Ruffino, F.; Canino, A.; Grimaldi, M.G.; Giannazzo, F.; Bongiorno, C.; Roccaforte, F.; Raineri, V. Self-organization of gold nanoclusters on hexagonal SiC and SiO_2 surfaces. *J. Appl. Phys.* **2007**, *101*, 064306. [CrossRef]

19. Ruffino, F.; Grimaldi, M.G. Formation of patterned arrays of Au nanoparticles on SiC surface by template confined dewetting of normal and oblique deposited nanoscale films. *Thin Solid Films* **2013**, *536*, 99–110. [CrossRef]
20. Ruffino, F.; Grimaldi, M.G. Dewetting of template-confined Au films on SiC surface: From patterned films to patterned arrays of nanoparticles. *Vacuum* **2014**, *99*, 28–37. [CrossRef]
21. Yang, L.; Zhao, H.; Fan, S.; Deng, S.; Lv, Q.; Lin, J.; Li, C.-P. Label-free electrochemical immunosensor based gold-silicon carbide nanocomposites for sensitive detection of human chorionic gonadotrophin. *Biosens. Bioelectron.* **2014**, *57*, 199–206. [CrossRef]
22. Zheng, G.; Xu, L.; Zou, X.; Liu, Y. Excitation of surface phonon polariton modes in gold gratings with silicon carbide substrate and their potential sensing applications. *Appl. Surf. Sci.* **2017**, *396*, 711–716. [CrossRef]
23. Zhigal'skii, G.P.; Jones, B.K. *The Physical Properties of Thin Metal Films*; Taylor and Francis: New York, NY, USA, 2003.
24. Barmak, K.; Coffey, K. *Metallic Films for Electronic, Optical and Magnetic Applications-Structure, Processing and Properties*; Woodhead Publishing: Cambridge, UK, 2014.
25. Johnston, R.L.; Wilcoxon, J.P. *Metal Nanoparticles and Nanoalloys*; Elsevier: Amsterdam, The Netherlands, 2012.
26. Xiong, Y.; Lu, X. *Metallic Nanostructures-From Controlled Synthesis to Applications*; Springer: New York, NY, USA, 2014.
27. Ruffino, F.; Crupi, I.; Simone, F.; Grimaldi, M.G. Formation and evolution of self-organized Au nanorings on indium-tin-oxide surface. *Appl. Phys. Lett.* **2011**, *98*, 023101. [CrossRef]
28. Sanzone, G.; Zimbone, M.; Cacciato, G.; Ruffino, F.; Carles, R.; Privitera, V.; Grimaldi, M.G. Ag/TiO_2 nanocomposite for visible light-driven photocatalysis. *Superlatt. Microstruct.* **2018**, *123*, 394–402. [CrossRef]
29. Gentile, A.; Ruffino, F.; Grimaldi, M.G. Complex-morphology metal-based nanostructures: Fabrication, characterization, and applications. *Nanomaterials* **2016**, *6*, 110. [CrossRef]
30. Ruffino, F.; Pugliara, A.; Carria, E.; Romano, L.; Bongiorno, C.; Fisicaro, G.; La Magna, A.; Spinella, C.; Grimaldi, M.G. Towards a laser fluence dependent nanostructuring of thin Au films on Si by nanosecond laser irradiation. *Appl. Surf. Sci.* **2012**, *258*, 9128–9137. [CrossRef]
31. Kim, K.-S.; Chung, G.-S. Characterization of porous cubic silicon carbide deposited with Pd and Pt nanoparticles as a hydrogen sensor. *Sens. Actuators B* **2011**, *157*, 482–487. [CrossRef]
32. Dai, H.; Chen, Y.; Lin, Y.; Xu, G.; Yang, C.; Tong, Y.; Guo, L.; Chen, G. A new metal electrocatalyst supported matrix: Palladium particles supported silicon carbide nanoparticles and its application for alcohol electrooxidation. *Electrochim. Acta* **2012**, *85*, 644–649. [CrossRef]
33. Zhang, Y.-W. *Bimetallic Nanostructures: Shape-controlled Synthesis for Catalysis, Plasmonics, and Sensing Applications*; Wiley: Hoboken, NJ, USA, 2018.
34. Yamauchi, M.; Kobayashi, H.; Kitagawa, H. Hydrogen storage mediated by Pd and Pt nanoparticles. *ChemPhysChem* **2009**, *10*, 2566–2576. [CrossRef] [PubMed]
35. Kundu, S.; Yi, S.-I.; Ma, L.; Chen, Y.; Dai, W.; Sinyukov, A.M.; Liang, H. Morphology dependent catalysis and enhanced Raman scattering (SERS) studies using Pd nanostructures in DNA, CTAB, and PVA scaffolds. *Dalton Trans.* **2017**, *46*, 9678–9691. [CrossRef] [PubMed]
36. Huang, H.; Jia, H.; Liu, Z.; Gao, P.; Zhao, J.; Luo, Z. Understanding of strain effects in the electrochemical reduction of CO_2: Using Pd nanostructures as an ideal platform. *Angew. Chem.* **2017**, *56*, 3594–3598. [CrossRef]
37. Ahmed, H.B.; Emam, H.E. Synergistic catalysis of monometallic (Ag, Au, Pd) and bimetallic (Ag–Au, Au–Pd) versus trimetallic (Ag–Au–Pd) nanostructures effloresced via analogical techniques. *J. Mol. Liq.* **2019**, *287*, 110975. [CrossRef]
38. Toshima, N.; Yonezawa, T. Bimetallic nanoparticles: Novel materials for chemical and physical applications. *New. J. Chem.* **1998**, *1998*, 1179–1201. [CrossRef]
39. Hao, C.-H.; Guo, X.-N.; Sankar, M.; Yang, H.; Ma, B.; Zhang, Y.-F.; Tong, X.-L.; Jin, G.-Q.; Guo, X.-Y. Synergistic effect of segregated Pd and Au nanoparticles on semiconducting SiC for efficient photocatalytic hydrogenation of nitroarenes. *ACS Appl. Mater. Interfaces* **2018**, *10*, 23029–23036. [CrossRef]
40. Zetterling, C.-M. *Process Technology for Silicon Carbide Devices*; INSPEC: London, UK, 2002.
41. Lundberg, N.; Ötling, M.; Tägtström, P.; Jansson, U. Chemical vapor deposition of tungsten Schottky diodes to 6H-SiC. *J. Electrochem. Soc.* **1996**, *143*, 1662–1667. [CrossRef]

42. Wolborski, M.; Bakowski, M.; Pore, V.; Ritala, M.; Leskelä, M.; Schöner, A.; Hallén, A. Characterization of aluminium and titanium oxides deposited on 4H-SiC by atomic layer deposition. *Mater. Sci. Forum* **2005**, *483–485*, 701–704. [CrossRef]
43. Arakawa, R.; Kawasaki, H. Functionalized nanoparticles and nanostructures surfaces for surface-assisted laser desorption/ionization mass spectrometry. *Anal. Sci.* **2010**, *26*, 1229–1240. [CrossRef]
44. Silina, Y.E.; Volmer, D.A. Nanostructured solid substrates for efficient laser desorption/ionization mass spectrometry (LDI-MS) of low molecular weight compounds. *Analyst* **2013**, *138*, 7053–7065. [CrossRef]
45. Gierz, I.; Suzuki, T.; Lee, D.S.; Krauss, B.; Riedl, C.; Starke, U.; Höchst, H.; Smet, J.H.; Ast, C.R.; Kern, K. Electronic decoupling of an epitaxial graphene monolayer by gold intercalation. *Phys. Rev. B* **2010**, *81*, 235408. [CrossRef]
46. Mishra, N.; Boeckl, J.; Motta, N.; Iacopi, F. Graphene growth on silicon carbide: A review. *Phys. Stat. Sol. A* **2016**, *213*, 2277–2289. [CrossRef]
47. Available online: http://www.genplot.com/doc/rump.htm (accessed on 12 April 2020).
48. Available online: https://www.ct.imm.cnr.it/?q=articles/characterizations-catania-unit (accessed on 12 April 2020).
49. Zinke-Allmang, M.; Feldman, L.C.; Grabov, M.H. Clustering on surfaces. *Surf. Sci. Rep.* **1992**, *16*, 377–463. [CrossRef]
50. Campbell, C.T. Ultrathin metal films and particles on oxide surfaces: Structural, electronic and chemisorptive properties. *Surf. Sci. Rep.* **1997**, *27*, 1–111. [CrossRef]
51. Venables, J.A.; Spiller, G.D.T.; Hanbucken, M. Nucleation and growth of thin films. *Rep. Prog. Phys.* **1984**, *47*, 399–459. [CrossRef]
52. Wang, Z.; Wynblatt, P. The equilibrium form of pure gold crystals. *Surf. Sci.* **1998**, *398*, 259–266. [CrossRef]
53. Yu, X.; Duxbury, P.M.; Jeffers, G.; Dubson, M.A. Coalescence and percolation in thin metal films. *Phys. Rev. B* **1991**, *44*, 13163(R)–13166(R). [CrossRef]
54. Heim, K.R.; Coyle, S.T.; Hembree, G.G.; Venables, J.A.; Scheinfein, M.R. Growth of nanometer-size metallic particles on CaF(111). *J. Appl. Phys.* **1996**, *80*, 1161–1170. [CrossRef]
55. Zhang, L.; Cosandey, F.; Persaud, R.; Madey, T.E. Initial growth and morphology of thin Au films on $TiO_2(110)$. *Surf. Sci.* **1999**, *439*, 73–85. [CrossRef]
56. Jeffers, G.; Dubson, M.A.; Duxbury, P.M. Island-to-percolation transition during growth of metal films. *J. Appl. Phys.* **1994**, *75*, 5016–5020. [CrossRef]
57. Duxbury, P.M.; Dubson, M.; Jeffers, G. Substrate inhomogeneity and the growth morphology of thin film. *Europhys. Lett.* **1994**, *26*, 601–606. [CrossRef]
58. Voss, R.F.; Laibowitz, R.B.; Alessandrini, E.I. Fractal (scaling) clusters in thin gold films near the percolation threshold. *Phys. Rev. Lett.* **1982**, *49*, 1441–1443. [CrossRef]
59. Fanfoni, M.; Tomellini, M.; Volpe, M. Coalescence and impingement between islands in thin film growth: Behavior of the island density kinetics. *Phys. Rev. B* **2001**, *64*, 075409. [CrossRef]
60. Vincent, R. A theoretical analysis and computer simulation of the growth of epitaxial films. *Proc. R. Soc. Lond.* **1971**, *321*, 53–68.
61. Briscoe, B.J.; Galvin, K.P. Growth with coalescence during condensation. *Phys. Rev. A* **1991**, *43*, 1906–1917. [CrossRef]

Article

Study of Inkjet-Printed Silver Films Based on Nanoparticles and Metal-Organic Decomposition Inks with Different Curing Methods

Peng Xiao [1,2], Yicong Zhou [2], Liao Gan [3], Zhipeng Pan [4], Jianwen Chen [1], Dongxiang Luo [5,*], Rihui Yao [2,6,*], Jianqiu Chen [2], Hongfu Liang [2] and Honglong Ning [2]

1 School of Physics and Optoelectronic Engineering, Foshan University, Foshan 528000, China; xiaopeng@fosu.edu.cn (P.X.); iamjwen@126.com (J.C.)
2 Institute of Polymer Optoelectronic Materials and Devices, State Key Laboratory of Luminescent Materials and Devices, South China University of Technology, Guangzhou 510640, China; zhou.yicong@mail.scut.edu.cn (Y.Z.); c.jianqiu@mail.scut.edu.cn (J.C.); 201530291429@mail.scut.edu.cn (H.L.); ninghl@scut.edu.cn (H.N.)
3 Air Force Representative Office in Zunyi District, Zunyi 563000, China; 15985282924@139.com
4 Guizhou Meiling Power Supply Co., Ltd., Zunyi 563000, China; pzpmlhg2004@126.com
5 School of Materials and Energy, Guangdong University of Technology, Guangzhou 510006, China
6 Guangdong Province Key Lab of Display Material and Technology, Sun Yat-sen University, Guangzhou 510275, China
* Correspondence: luodx@gdut.edu.cn (D.L.); yaorihui@scut.edu.cn (R.Y.)

Received: 23 June 2020; Accepted: 11 July 2020; Published: 12 July 2020

Abstract: Currently, inkjet printing conductive films have attracted more and more attention in the field of electronic device. Here, the inkjet-printed silver thin films based on nanoparticles (NP) ink and metal-organic decomposition (MOD) ink were cured by the UV curing method and heat curing method. We not only compared the electrical resistivity and adhesion strength of two types of silver films, but also studied the effect of different curing methods on silver films. The silver films based on NP ink had good adhesion strength with a lowest electrical resistivity of 3.7×10^{-8} Ω·m. However, the silver film based on MOD ink had terrible adhesion strength with a lowest electrical resistivity of 2×10^{-8} Ω·m. Furthermore, we found a simple way to improve the terrible adhesion strength of silver films based on MOD ink and tried to figure out the mechanisms. This work offers a further understanding of the different performances of two types of silver films with different curing methods.

Keywords: inkjet printing; nanoparticle; metal-organic decomposition; silver thin film; adhesion strength; electrical resistivity

1. Introduction

Inkjet printing has been used to print documents in offices for about 50 years. Now, people are aware of its advantages in the electronic field, such as direct patterning without masks, micrometer resolution, low cost, non-contacting printing, etc. [1,2]. Therefore, more and more attentions are paid to the applications of inkjet printing conductive films in electronic products, such as radio frequency identification (RFID) tag [3], thin film transistor [4–6], solar cell [7], OLED [8,9], etc. The performance of conductive films mainly relies on the composition of materials and post treatments. Currently, conductive materials widely used in inkjet printing conductive films are silver based inks because silver is more chemically stable than copper and cheaper than gold. In addition, the melting temperature of silver nanoparticles is much lower than bulk silver and decreases as the size of particles decreases, which is beneficial to the low-temperature manufacture [10,11]. So far, silver-based inks mainly include nanoparticles (NP) silver ink and metal-organic decomposition (MOD) silver ink. The NP silver ink consists of silver nanoparticles, organic solvents, dispersants, organic coating surrounding silver particles, and other organic additives, while the MOD silver ink is mainly composed of silver-based precursors and organic solvents [12]. For the NP silver ink, Alessandro et al. [13] realized a power amplifier on a rigid alumina substrate by adopting a hybrid Ag-based ink, whose electrical properties were comparable to that of pure-metallic ink systems. Dearden et al. [14] reported a low curing temperature silver ink for inkjet printing silver films. Zhao et al. [15] improved the electrical conductivity of printed silver films by using a novel carbon nanotubes/silver NP ink. For the MOD silver ink, Nie et al. [16] printed the silver conductive film based on a silver citrate conductive ink. Kalio et al. [17] developed a lead-free silver ink for the front side metallization of silicon solar cells. In a word, most previous works [18–21] only focused on the effects of heat curing on a single silver ink or printing silver films. However, there are few reports on comparing different types of silver inks. Therefore, it is necessary to figure out different features of NP silver ink and MOD silver ink, which can help researchers find their different scope of applications. The posttreatment methods used in conductive films mainly include heat curing, laser sintering [22], electric sintering [23,24], microwave sintering [3], etc. As a low temperature and fast curing method, UV curing was promising in flexible electronics. Therefore, it was also meaningful to study the effects of UV curing on the NP and MOD silver inks.

In this work, NP silver ink and MOD silver ink were chosen to prepare the inkjet-printed silver thin films on glass substrates. UV curing and the traditional heat curing method were used to cure silver films to investigate the effect of different curing methods on silver films. We investigated the effects of two different curing methods on electrical resistivity and adhesion strength of two types of silver films. Furthermore, we also tried to figure out the different mechanisms that two types of silver films were bonded to the glass substrate under different curing methods.

2. Materials and Methods

In our work, the conventional alkali-free glass with a thickness of 0.7 mm was used as the substrates. To remove dust and contamination on the surface, the substrates were ultrasonicated in isopropyl alcohol, tetrahydrofuran, deionized water, and isopropyl alcohol in sequence. Then, the clean substrates were dried for use without an additional pretreatment. The NP silver ink used in inkjet printing was DGP-40LT-15C (purchased from Advanced Nano Products Co. Ltd., Sejong-si, Korea), while the MOD silver ink was TEC-IJ-010 (purchased from Inktec Co. Ltd., Ansan, Japan). A dimatix material printer (DMP-2800) with a 10 pL cartridge was used to print silver films. During the printing, the inks were printed onto the substrates with a drop spacing of 35 μm. In addition, the substrate temperature of the printer was set at 30 and 50 °C in order to get continous silver films based on NP ink and MOD ink, respectively. After the printing, the silver films were cured by UV curing or heat curing on a hot plate under ambient conditions. The UV light curing system used in UV curing was IntelliRay UV0832 (Uvitron International Inc, West Springfield, MA, USA) and the power of UV lamp in the system is 600 W.

The electrical resistivity of the films was calculated from the equation of $\rho = R_S \times h$, where ρ is the electrical resistivity, R_s is the sheet resistance, and h is the thickness of films. The sheet resistance and thickness were measured by a digital four-probe tester (KDY-1, Guangzhou Kunde Co. Ltd., Guangzhou, China) and a step profiler (Dektak, Veeco, Plainview, NY, USA), respectively. A scanning electron microscopy (SEM, NOVA NANOSEM 430, FEI, Hillsboro, OR, USA) with an energy dispersive X-ray spectrometer (EDS) was used to obtain a relative element content, surface information, and interface information between silver films and substrates. In addition, the adhesion strength of silver films was tested by the tape test in which we printed 64 square silver films with an edge length of 0.7 mm onto the substrates and used the 3M tape to test the adhesion strength.

3. Results and Discussion

Figure 1a,b shows the influence of annealing temperature on the electrical resistivity of two types of silver films, in which the sample in 25 °C was dried naturally without additional curing and just prepared for comparison (not shown). A digital four-probe tester was used to measure sheet resistance and the electrical resistivity of the films was calculated from the equation of $\rho = R_S \times h$. The silver film in 25 °C was not conductive (not shown), indicating that a large number of organic coatings existed in the silver films. As shown in Figure 1a,b, the lowest electrical resistivity of the silver films based on NP ink (3.7×10^{-8} Ω·m) was larger than that of the silver films based on MOD ink (2.6×10^{-8} Ω·m). The electrical resistivity of two types of silver films both decreased dramatically with the increasing temperature at first because of the evaporation of solvent which was consistent with the decreasing relative carbon content, as shown in Figure 1c. However, when the temperature continued to increase, the electrical resistivity increased slightly. Compared with Figure 2a,c or Figure 2d,f, there were more pores with a large size on the surface of silver films in 200 °C. It indicated that the faster evaporation rate caused more pores in silver films, and a faster curing rate made the silver inks solidify more quickly so that less pores could be eliminated by the flow of liquid inks, which caused a worse mutual contact between the silver particles. On the other hand, the rapid solidification of silver ink resulted in a small amount of organic substances trapped inside silver films, which was consistent with a slightly increasing relative carbon content in 200 °C (Figure 1c). As a result, the electrical resistivity of silver films increased. In addition, we speculated that more pores with a large size existing on the surface of silver film based on NP silver ink than that based on MOD silver ink was due to the removal of organic coatings in 200 °C.

The temperature of the lowest electrical resistivity of the silver films based on NP ink was smaller than that based on MOD ink. It may be ascribed to the existence of the organic coating surrounding the silver particles in the NP ink. The organic coating could not only prevent silver nanoparticles from agglomeration when the silver ink was not cured, but also serve as the binder which leads to a good mutual contact between the silver particles. Therefore, when the solvent was evaporated, organic coating made the silver particle connected with each other so that the temperature (140 °C) of the lowest electrical resistivity of the silver films based on NP ink was lower than that (180 °C) based on MOD ink. However, the existence of organic coating made the lowest electrical resistivity of silver films based on the NP ink larger than that based on the MOD ink. What is more, when organic coating was removed in high temperature, it caused extra pores in the silver films. This explained the phenomenon that the electrical resistivity of the silver films based on NP ink increased slightly in high temperature (Figure 1a). As shown in Figure 2c,f, larger pores were observed in the silver films based on NP ink in 200 °C than that based on MOD ink, which can be ascribed to the removal of organic coating. Compared with NP ink, the silver particles from the MOD ink were not surrounded by organic coating so that higher temperatures were needed to melt silver particles to form the mutual connection between adjacent particles for a lower electrical resistivity. Therefore, less curing time was needed to remove the organic residues in silver films based on MOD ink, which explained why the curing time of the silver films based on MOD ink was 10 min instead of 30 min. As shown in Figure 2e, a large amount of silver particles in silver films based on MOD ink were not connected to the adjacent particles, while the connection between silver particles in silver films based on NP ink in 140 °C was closer (Figure 2b). Therefore, it was necessary to increase the curing temperature to attain the lower electrical resistivity for silver films based on MOD ink.

Figure 3a shows the electrical resistivity of silver films based on MOD ink with different UV curing conditions. The electrical resistivity of silver films decreased with the decreasing D (distance between UV lamp and silver films), which was consistent with the decreasing carbon content shown in Figure 3b. In addition, the electrical resistivity decreased with the increase of curing time. The silver film with a low electrical resistivity of ~2.03×10^{-8} Ω·m was obtained at D = 23 cm for 480 s by UV curing. Furthermore, the electrical resistivity (~2.03×10^{-8} Ω·m) of silver films with UV curing was smaller than the heat-cured one (2.63×10^{-8} Ω·m, in Figure 1b), which can be ascribed to the low temperature and good thermal uniformity for the UV curving method. Since less pores were formed in low temperature, the lowest electrical resistivity of silver films with UV curing was lower than the heat-cured one. However, the silver film with UV curing at D = 37 cm for 270 s was not conductive in Figure 3a. As shown in Figure 4a, the silver film with UV curing at D = 37 cm for 270 s was discontinuous, leading to the inferior conductivity. As D decreased and the curing time increased, the silver particles in silver films started to melt and merge into bigger particles as shown in Figure 4b,c. As a result, the electrical resistivity of silver films became smaller. As shown in Table 1, the resistivity value (2.03×10^{-8} Ω·m) in our work is lower than that in the previous work, and the UV curing shows a great potential in the field of flexible electronics which need a low temperature process.

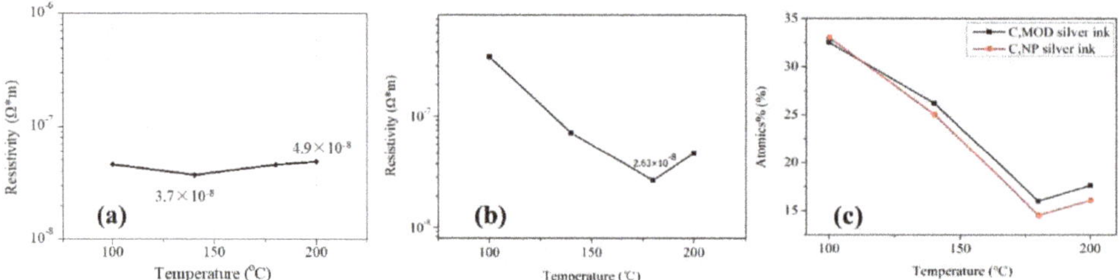

Figure 1. Electrical resistivity and relative carbon content of silver film at different heat-cured temperatures: (**a**) Silver films based on the nanoparticles (NP) ink were heat cured for 30 min; (**b**) silver films based on the metal-organic decomposition (MOD) ink were heat cured for 10 min; (**c**) relative carbon content of heat-cured silver films.

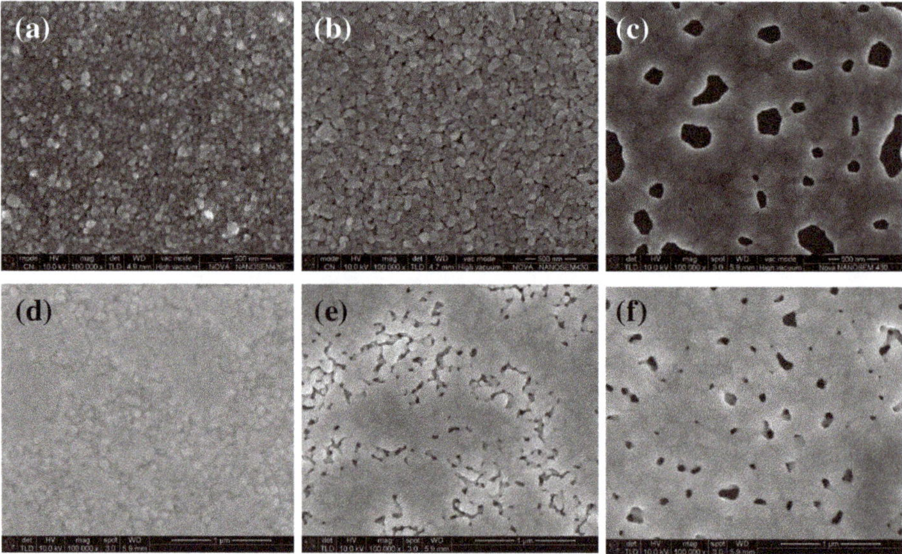

Figure 2. SEM images of heat-cured silver films: Silver films based on the NP ink were cured in (**a**) 100, (**b**) 140, and (**c**) 200 °C for 30 min, respectively; silver films based on the MOD ink were cured in (**d**) 100, (**e**) 140, and (**f**) 200 °C for 10 min, respectively.

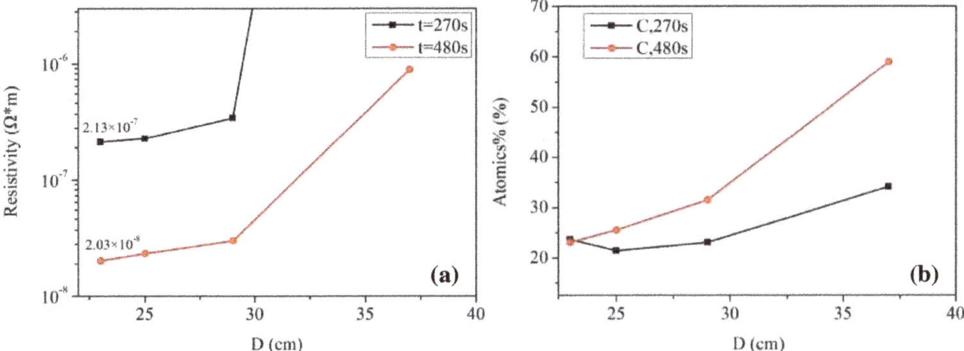

Figure 3. Electrical resistivity and carbon content of ultra violet (UV)-cured silver films based on MOD ink: (**a**) The electrical resistivity of silver films; (**b**) relative carbon content of silver films. D was defined as the distance between the samples and UV lamp.

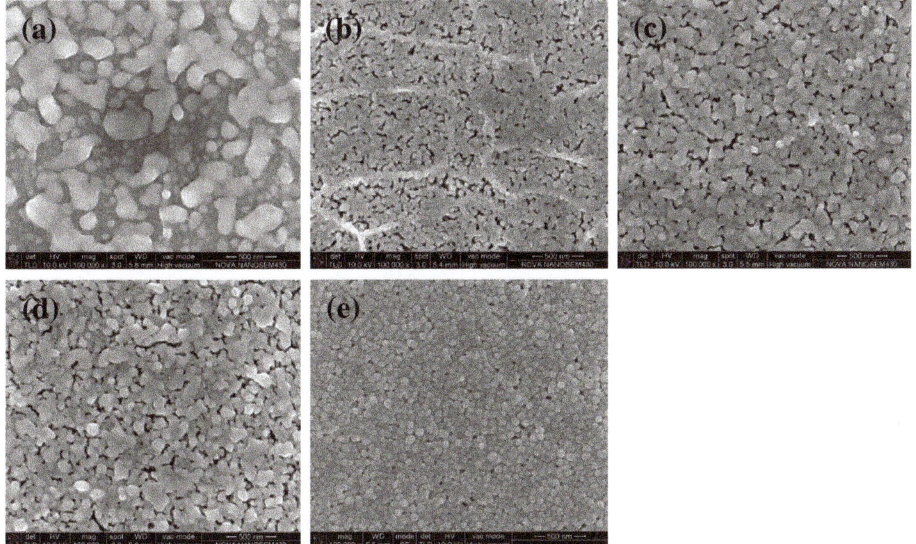

Figure 4. SEM images of silver films with UV curing at: (**a**) D = 37 cm for 270 s; (**b**) D = 23 cm for 270 s; (**c**) D = 23 cm for 480 s; (**d**) D = 25 cm for 480 s; (**e**) D = 25 cm for 480 s [20]. (**a**–**d**) were the films based on MOD ink while (**e**) was the film based on NP ink.

Table 1. Comparison of resistance of silver films between the previous work and this work.

Ref.	Ink Type	Preparation	Post Treatment	Electrical Resistivity/(Ω·m)
2	MOD	Inkjet printing	heat curing (170 °C)	8.4×10^{-8}
13	NPs	Inkjet printing	heat curing (150 °C)	$(3–4) \times 10^{-7}$
14	NPs	Inkjet printing	heat curing (150 °C)	1.8×10^{-7}
20	NPs	Inkjet printing	UV curing	6.69×10^{-8}
This work	MOD	Inkjet printing	UV curing	2.03×10^{-8}

Our previous work [25] had studied the effects of UV curing on the electrical resistivity of silver films based on NP ink, and the silver film with a low electrical resistivity of ~6.7×10^{-8} Ω·m was obtained at D = 25 cm for 480 s. The electrical resistivity of the silver film based on NP ink was about 2.5×10^{-7} Ω·m when it was cured by UV at D = 37 cm for 180 s. By contrast, when D = 37 cm, the silver film based on MOD ink was not conductive until the UV curing time was increased from 270 to 480 s, as shown in Figure 3a. For the precursor in MOD, the process of decomposing and forming silver particles took a lot of energy (UV irradiation: 600 W, D = 23 cm for 480 s), so it needed plenty of time (~480 s) to form the conductive film when D was large. However, the electrical resistivity of the silver film based on MOD ink was much smaller than that of the silver film based on NP ink when they were UV cured at D = 25 cm for 480 s. The relatively large resistivity of the silver film based on NP ink can be ascribed to the organic coating surrounding silver particles. Since the organic coating could prevent silver particles from merging into bigger particles, and UV curing could not remove it completely because of the relatively low energy. As shown in Figure 4e, the silver particles were closely packed but they did not melt and merge into bigger particles. However, the silver particles from the MOD silver ink were not surrounded by organic coating, so they were able to melt and merge into big particles in low temperature (Figure 4d). Therefore, the lowest electrical resistivity of the silver film based on MOD ink (~2.03×10^{-8} Ω·m) was much lower than that based on NP ink when they were UV cured at D = 25 cm for 480 s.

The adhesion strength of silver films on the glass substrate were characterized by a tape test according to the ASTM international standard [26]. The test steps are as follows: First, an 8 × 8 silver electrodes array with an edge length of 0.7 mm was inkjet printed on the glass substrate and cured in succession. Then, we put the tape on the surface of the silver electrodes and extruded the bubbles in the tape. After resting in 90 s, we tore the tape at a 180° angle from the substrate and observed the falling off of the silver electrodes. Figure 5 shows the adhesion strength of silver films based on NP ink. We found that silver films based on NP ink had good adhesion strength on the glass substrate and almost all of the square films were not detached from the substrates. The adhesion strength of silver electrodes on the glass substrate was about 4–5 B according to the ASTM international standard [26]. We speculated that the remaining organic coating surrounding the silver particles may bond a silver film with the substrate. To figure out the mechanism, we observed SEM images of the cross section of silver films shown in Figure 6. Since the organic coating in NP silver ink usually has a relatively high boiling point and will not be removed in low temperature, the remaining organic coating may serve as the binder [27] and the interfacial pores can be filled with the coating, improving the adhesion strength of silver films, as shown in Figure 7a. As shown in Figure 6, silver particles in the interface between the substrates and films cured by two curing methods had a really close contact with the substrates. What is more, compared with the silver film cured in 140 °C (Figure 2b,e), there were no apparent edges of particles inside the UV-cured silver film and few pores in the interface. It indicated that the silver particles were surrounded by organic coating in low temperature. Therefore, the organic coating bonded silver film with the substrates, leading to the good adhesion strength of silver films is based on NP ink.

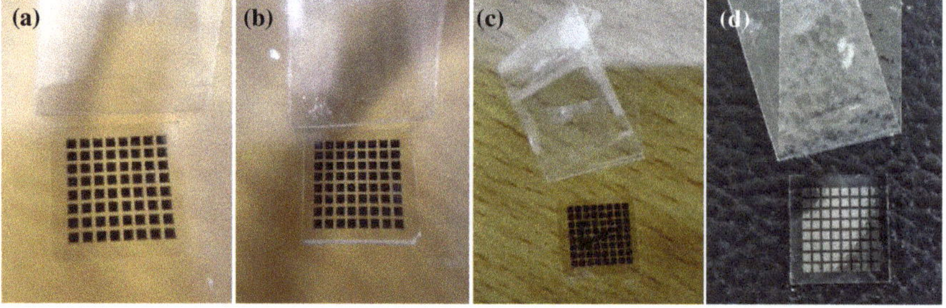

Figure 5. Photo of silver films based on NP ink after the tape test: (**a**) Heat cured in 80 °C for 30 min; (**b**) heat cured in 140 °C for 30 min; (**c**) UV cured at D = 37 cm for 180 s; (**d**) UV cured at D = 25 cm for 480 s.

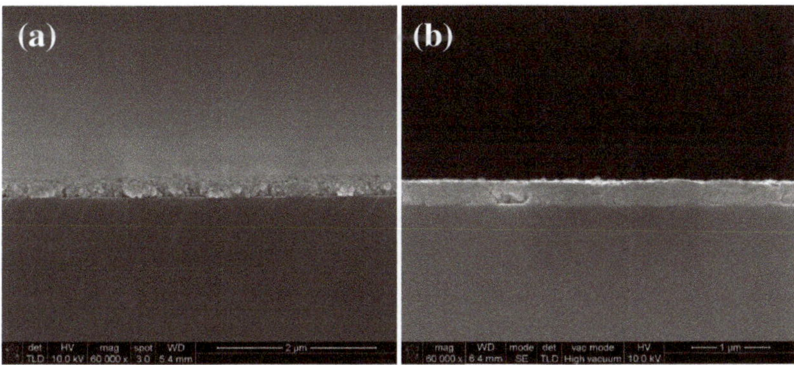

Figure 6. SEM images of the cross section of silver films based on NP ink: (**a**) Heat cured in 140 °C for 30 min; (**b**) UV cured at D = 25 cm for 480 s.

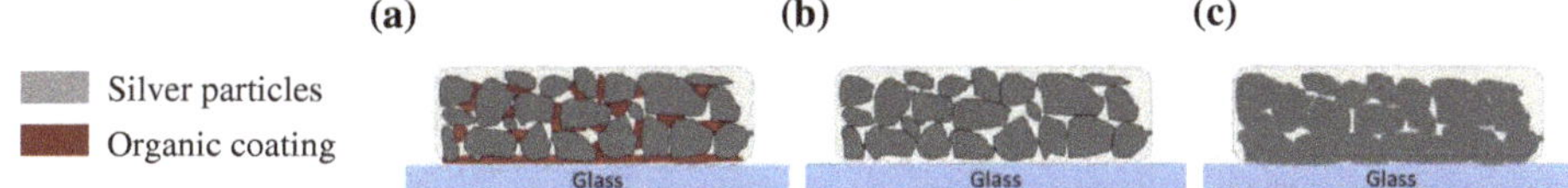

Figure 7. Schematic of the possible adhesion mechanism of (**a**) silver films based on NP ink; (**b**) silver films based on MOD ink cured in low temperature; (**c**) silver films based on MOD ink cured in high temperature.

Without the existence of organic coating, the adhesion strength of silver films based on MOD ink was supposed to be worse than that based on NP ink. Figure 8 shows the adhesion strength of silver films based on MOD ink. When silver films were cured by UV curing or heated at 180 °C, the adhesion strength was poor and most of the square films were detached from the substrates, which was consistent with our assumption. Since the solvent could be removed in a really low temperature and there was no organic coating in the MOD silver ink, the connection between silver films and substrates was poor. However, when silver films were heat cured, the adhesion strength increased with the increasing temperature. When the silver film was heat cured at 200 °C for 10 min, few square films were detached from the substrate, indicating that the mechanism of connecting silver films based on MOD ink with the substrates was probably different from that based on NP ink.

As shown in Figure 9a, many pores existed in the interface and most of the silver particles were not melted and merged into big particles when the film was cured in 160 °C. As the temperature increased to 180 °C, the size of the pores in the interface became smaller due to the melting of the particles shown in Figure 9b. Without the existence of the organic coating, the pores in the interface of the silver films based on MOD ink led to a poor connection between the films and substrates, as shown in Figure 7b. Consequently, the adhesion strength of the silver films cured in relatively low temperature was relatively poor. As shown in Figure 9c, when the film was cured at 200 °C, all silver particles were melted and no apparent pores were found in the interface, which indicated that the contacting area between the silver film and the substrate became larger and the connection between the film and the substrate became better. Therefore, the amount of the pores in the interface can be reduced by melting the particles in high temperature and a few pores meant a good connection between the films and substrates, leading to the good adhesion strength [28,29]. In addition, we speculated that a higher temperature could probably promote the mutual diffusion between silver films and substrates [30], which was needed to be proved in the future work.

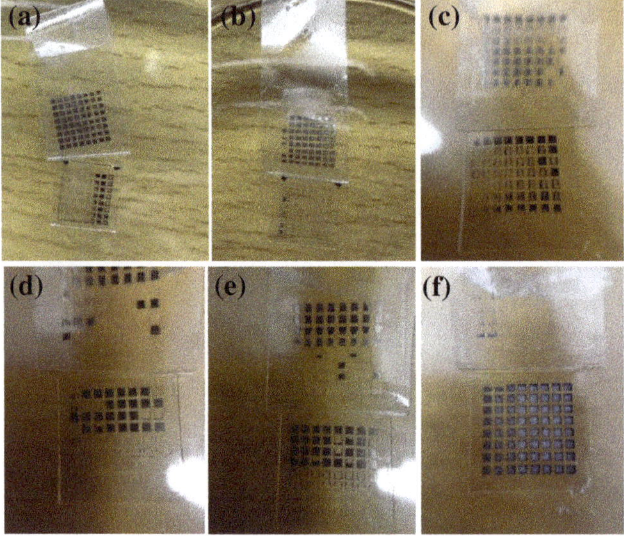

Figure 8. Photos of silver films based on MOD ink after the tape test: (a) UV cured at D = 37 cm for 480 s; (b) UV cured at D = 29 cm for 480 s; (c) UV cured at D = 23 cm for 480 s; (d) heat cured at 160 °C for 10 min; (e) heat cured at 180 °C for 10 min; (f) heat cured at 200 °C for 10 min.

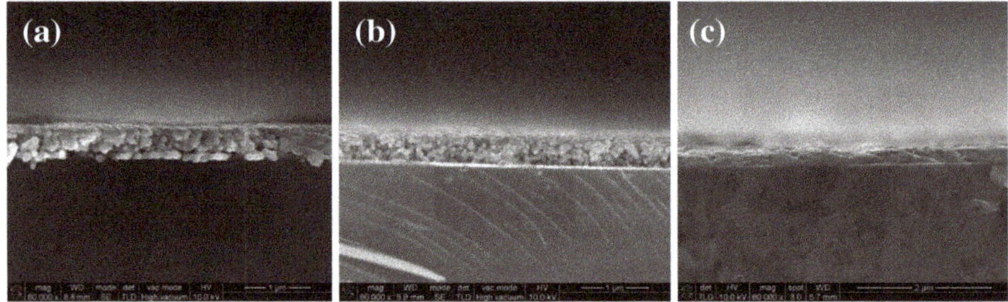

Figure 9. SEM images of cross section of heat-cured silver films based on MOD ink at: (a) 160 °C; (b) 180 °C; (c) 200 °C for 10 min.

To verify our hypothesis, the silver films prepared by DC magnetron sputtering were used for comparison to eliminate the interference of the residue organic substance. The silver films were cured on hot plate at a temperature of 100 and 200 °C for 10 min, respectively. According to the results of the tape test shown in Figure 10, the adhesion strength of the silver film cured at 100 °C was really poor. However, when the temperature increased to 200 °C, the adhesion strength of the silver film was improved significantly. As shown in Figure 10c, the silver particles were melted and no apparent pores existed in the interface, which was similar to Figure 9c. It indicated that the connection between the silver films and glass substrates could be enhanced by increasing the curing temperature without the help of organic coating. Correspondingly, the adhesion strength increased with the increase of curing temperature. By comparing Figure 8f with Figure 10b, the adhesion strength of the silver film fabricated by DC magnetron sputtering was still worse than that of the inkjet-printed silver film based on MOD ink, indicating that there were possibly some residual organic substances on the interface between the substrate and silver film based on MOD ink. Therefore, the adhesion strength of silver film based on MOD ink was mainly affected by the curing temperature (or pores in the interface) and residual organics.

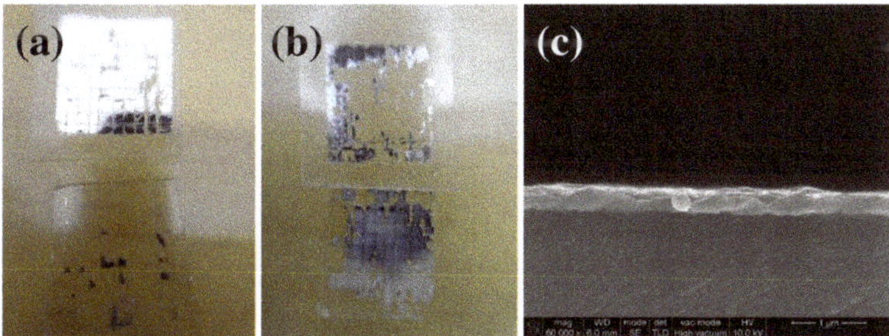

Figure 10. Photos of silver films deposited by DC magnetron sputtering after the tape test: (**a**) Heat cured at 100 °C for 10 min; (**b**) heat cured at 200 °C for 10 min; (**c**) SEM image of cross section of silver films heat cured at 200 °C for 10 min.

4. Conclusions

In conclusion, the NP silver ink and MOD silver ink were used to prepare silver films which were cured by UV curing or the heat curing method, and the effects of two curing methods on the electrical resistivity and adhesion strength of two types of silver films were systematically investigated. The silver films based on NP ink had worse conductivity than that based on MOD ink due to the existence of organic coating in NP ink. In our work, for silver films based on NP ink, the lowest electrical resistivity of the UV-cured films was almost two times larger than that of the heat-cured films because the organic coating stopped silver particles from merging into big particles in low temperature. For silver films based on MOD ink, the lowest electrical resistivity of the UV-cured films ($\sim 2.03 \times 10^{-8}$ Ω·m) was similar to that of the heat-cured films (2.6×10^{-8} Ω·m), which means that silver films based on MOD ink could attain a low electrical resistivity in the low-temperature process. However, silver films based on MOD ink had to be cured in a relatively high temperature to attain good adhesion strength while the silver films based on NP ink could attain good adhesion strength in a relatively low temperature with the help of organic coating. Furthermore, the existence of the remaining organic coating may cause interface problems of multilayers' structure. These results exhibited that our work could offer a further understanding of the different features of inkjet-printed silver thin films based on NP ink and MOD ink and help other researchers choose proper conductive materials and curing methods to fabricate different electronic devices. Next, we plan to study the flexible electronic devices based on printing silver electrodes in the future.

Author Contributions: P.X., Y.Z., R.Y., D.L., and H.N. conceived the idea; P.X. and Y.Z. wrote the paper; L.G., Z.P., J.C. (Jianwen Chen), J.C. (Jianqiu Chen), and H.L. advised the paper. All authors reviewed the paper. All authors have read and agreed to the published version of the manuscript.

Funding: This research was funded by the National Natural Science Foundation of China (grant no. 61804029 and 51771074), the Guangdong Natural Science Foundation (no. 2018A030310353 and 2017A030310028), the Guangdong Science and Technology Project (no. 2016B090907001 and 2019B010934001), the Science and Technology Project of Guangzhou (no. 201904010344), the Foundation for Young Talents in Higher Education of Guangdong (grant no. 2018KQNCX275), the Fundamental Research Funds for the Central Universities (no. 2019MS012), the Open Project of Guangdong Province Key Lab of Display Material and Technology (no. 2017B030314031), the Engineering Discipline Open Fund of Foshan University, and the Project of Foshan Education Bureau (2019XJZZ02).

Conflicts of Interest: The authors declare no conflict of interest.

References

1. Mustonen, T.; Kordás, T.; Saukko, S.; Toth, G.; Penttilla, J.S.; Helisto, P.; Seppa, H.; Jantunen, H. Inkjet printing of transparent and conductive patterns of single-walled carbon nanotubes and PEDOT-PSS composites. *Phys. Status Solidi (b)* **2007**, *244*, 4336–4340. [CrossRef]
2. Dong, Y.; Li, X.; Liu, S.; Zhu, Q.; Li, J.G.; Sun, X.D. Facile synthesis of high silver content MOD ink by using silver oxalate precursor for inkjet printing applications. *Thin Solid Films* **2015**, *589*, 381–387. [CrossRef]

3. Perelaer, J.; Klokkenburg, M.; Hendrik, C.E.; Schubert, U.S. Microwave flash sintering of inkjet-printed silver tracks on polymer substrates. *Adv. Mater.* **2009**, *21*, 4830–4834. [CrossRef] [PubMed]
4. Yang, C.; Fang, Z.; Ning, H.; Tao, R.; Chen, J.; Zhou, Y.; Zheng, Z.; Yao, R.; Wang, L.; Peng, J. Amorphous InGaZnO thin film transistor fabricated with printed silver salt ink source/drain electrodes. *Appl. Sci.* **2017**, *7*, 844. [CrossRef]
5. Ning, H.; Chen, J.; Fang, Z.; Tao, R.; Cai, W.; Yao, R.; Hu, S.; Zhu, Z.; Zhou, Y.; Yang, C.; et al. Direct inkjet printing of silver source/drain electrodes on an amorphous InGaZnO layer for thin-film transistors. *Materials* **2017**, *10*, 51. [CrossRef]
6. Noguchi, Y.; Sekitani, T.; Yokota, T.; Someya, T. Direct inkjet printing of silver electrodes on organic semiconductors for thin-film transistors with top contact geometry. *Appl. Phys. Lett.* **2018**, *93*, 43303. [CrossRef]
7. Shin, D. Fabrication of an inkjet-printed seed pattern with silver nanoparticulate ink on a textured silicon solar cell wafer. *J. Micromech. Microeng.* **2010**, *20*, 125003. [CrossRef]
8. Shu, Z.; Beckert, E.; Eberhardt, R.; Tuennermann, A. ITO-free, inkjet-printed transparent organic light-emitting diodes with a single inkjet-printed Al: ZnO: PEI interlayer for sensing applications. *J. Mater. Chem. C* **2017**, *5*, 11590–11597. [CrossRef]
9. Ye, T.; Jun, L.; Kun, L.; Hu, W.; Ping, C.; Ya-Hui, D.; Zheng, C.; Yun-Fei, L.; Hao-Ran, W.; Yu, D. Inkjet-printed Ag grid combined with Ag nanowires to form a transparent hybrid electrode for organic electronics. *Org. Electron.* **2017**, *41*, 179–185. [CrossRef]
10. Allen, G.L.; Bayles, R.A.; Gile, W.W.; Jesser, W.A. Small particle melting of pure metals. *Thin Solid Films* **1986**, *144*, 297–308. [CrossRef]
11. Jiang, Q.; Zhang, S.; Zhao, M. Size-dependent melting point of noble metals. *Mater. Chem. Phys.* **2003**, *82*, 225–227. [CrossRef]
12. Kamyshny, A. Metal-based inkjet inks for printed electronics. *Open Appl. Phys. J.* **2011**, *4*, 19–36. [CrossRef]
13. Chiolerio, A.; Camarchia, V.; Quaglia, R.; Pirola, M.; Pandolfi, P.; Pirri, C.F. Hybrid Ag-based inks for nanocomposite inkjet printed lines: RF properties. *J. Alloys Compd.* **2015**, *615*, S501–S504. [CrossRef]
14. Dearden, A.L.; Smith, P.J.; Shin, D.Y.; Reis, N.; Derby, B.; O'Brien, P. A low curing temperature silver ink for use in ink-jet printing and subsequent production of conductive tracks. *Macromol. Rapid Commun.* **2005**, *26*, 315–318. [CrossRef]
15. Zhao, D.; Liu, T.; Park, J.G.; Zhang, M.; Chen, J.-M.; Wang, B. Conductivity enhancement of aerosol-jet printed electronics by using silver nanoparticles ink with carbon nanotubes. *Microelectron. Eng.* **2012**, *96*, 71–75. [CrossRef]
16. Nie, X.; Wang, H.; Zou, J. Inkjet printing of silver citrate conductive ink on PET substrate. *Appl. Surf. Sci.* **2012**, *261*, 554–560. [CrossRef]
17. Kalio, A.; Leibinger, M.; Filipovic, A.; Krüger, K.; Glatthaar, M. Development of lead-free silver ink for front contact metallization. *Sol. Energy Mater. Sol. Cells* **2012**, *106*, 51–54. [CrossRef]
18. Yang, X.; He, W.; Wang, S.; Zhou, G.; Tang, Y. Preparation of high-performance conductive ink with silver nanoparticles and nanoplates for fabricating conductive films. *Mater. Manuf. Process.* **2012**, *28*, 1–4. [CrossRef]
19. Chang, Y.; Wang, D.; Tai, Y.; Yang, Z. Preparation, characterization and reaction mechanism of a novel silver-organic conductive ink. *J. Mater. Chem.* **2012**, *22*, 25296–25301. [CrossRef]
20. Lee, C.; Dupeux, M.; Tuan, W. Influence of firing temperature on interface adhesion between screen-printed Ag film and $BaTiO_3$ substrate. *Mater. Sci. Eng. A* **2007**, *467*, 125–131. [CrossRef]
21. Kan-Sen, C.; Kuo-Cheng, H.; Hsien-Hsuen, L. Fabrication and sintering effect on the morphologies and conductivity of nano-Ag particle films by the spin coating method. *Nanotechnology* **2005**, *16*, 779–784.
22. Maekawa, K.; Yamasaki, K.; Niizeki, T.; Mita, M.; Matsuba, Y.; Terada, N.; Saito, H. Drop-on-demand laser sintering with silver nanoparticles for electronics packaging. *IEEE Trans. Compon. Packag. Manuf.* **2012**, *2*, 868–877. [CrossRef]
23. Hummelgard, M.; Zhang, R.; Nilsson, H.; Olin, H. Electrical sintering of silver nanoparticle ink studied by in-situ TEM probing. *PLoS ONE* **2011**, *6*, e17209. [CrossRef] [PubMed]
24. Allen, M.L.; Aronniemi, M.; Mattila, T.; Alastalo, A.; Ojanpera, K.; Suhonen, M.; Seppa, H. Electrical sintering of nanoparticle structures. *Nanotechnology* **2008**, *19*, 175201. [CrossRef] [PubMed]

25. Ning, H.; Zhou, Y.; Fang, Z.; Yao, R.; Tao, R.; Chen, J.; Cai, W.; Zhu, Z.; Yang, C.; Wei, J.; et al. UV-cured inkjet-printed silver gate electrode with low electrical resistivity. *Nanoscale Res. Lett.* **2017**, *12*, 546. [CrossRef] [PubMed]
26. *ASTM D3359-97. Standard Test Methods for Measuring Adhesion by Tape Test; ASTM International: West Conshohocken,* PA, USA, 1997. [CrossRef]
27. Kim, J.; Kim, K.; Jang, K.; Jung, S.; Kim, T. Enhancing adhesion of screen-printed silver nanopaste films. *Adv. Mater. Interfaces* **2015**, *2*, 1500283. [CrossRef]
28. Lee, C.; Dupeux, M.; Tuan, W. Influence of bulk and interface porosity on the adhesion of sintered Ag films on barium titanate substrates. *Adv. Eng. Mater.* **2011**, *13*, 64–67. [CrossRef]
29. Joo, S.; Baldwin, D.F. Adhesion mechanisms of nanoparticle silver to substrate materials: Identification. *Nanotechnology* **2010**, *21*, 55204. [CrossRef]
30. Wu, X.; Shao, S.; Chen, Z.; Cui, Z. Printed highly conductive Cu films with strong adhesion enabled by low-energy photonic sintering on low-tg flexible plastic substrate. *Nanotechnology* **2017**, *28*, 35203. [CrossRef]

Article

Influence of Illumination on Porous Silicon Formed by Photo-Assisted Etching of p-Type Si with a Different Doping Level

Olga Volovlikova *, Sergey Gavrilov and Petr Lazarenko

Institute of Advanced Materials and Technologies, National Research University of Electronic Technology (MIET), Moscow 124498, Russia; pcfme@miee.ru (S.G.); aka.jum@gmail.com (P.L.)
* Correspondence: 5ilova87@gmail.com; Tel.: +7-903-292-1507

Received: 9 January 2020; Accepted: 12 February 2020; Published: 14 February 2020

Abstract: The influence of illumination intensity and p-type silicon doping level on the dissolution rate of Si and total current by photo-assisted etching was studied. The impact of etching duration, illumination intensity, and wafer doping level on the etching process was investigated using scanning electron microscopy (SEM), atomic force microscopy (AFM), and Ultraviolet-Visible Spectroscopy (UV–Vis–NIR). The silicon dissolution rate was found to be directly proportional to the illumination intensity and inversely proportional to the wafer resistivity. High light intensity during etching treatment led to increased total current on the Si surface. It was shown that porous silicon of different thicknesses, pore diameters, and porosities can be effectively fabricated by photo-assisted etching on a Si surface without external bias or metals.

Keywords: photo-assisted etching; porous silicon; illumination; doping level; total current; reflectance

---

1. Introduction

Porous silicon (por-Si) has been receiving a great deal of attention due to its interesting physical and optical [1] properties and promising potential technological applications in the fields of sensing [2,3], Li-ion batteries [4], and solar cells [5]. The morphology of the silicon wafer may be modified by electrochemical wet etching [6–8], galvanic etching [9–11], or metal-assisted chemical etching (MACE) [12,13] to produce pores, nanowires, and microstructures. These methods have already been studied in detail. A cheap, simple way to make porous silicon layers without surface contamination is important for silicon technology. It is possible to produce porous silicon layers in solutions containing HF and H_2O_2 under illumination without external bias or a metal film on the Si surface. This technique is the photo-assisted etching of Si (PhACE). The silicon wafer is not connected to the external power supply, and the electric field is due to band bending. PhACE can produce porous silicon for solar cell applications; however, this method is poorly understood. The physical properties of porous silicon are determined by two large groups of factors that affect carrier density on the surface. The first group of factors includes doping type and charge carrier concentration [10,14]. The second group includes illumination during etching [15,16]. The influence of lamp illumination on the properties of n-type porous silicon samples during the etching process has been deeply investigated. For example, Koker and Kolasinski [17] investigated laser-assisted porous silicon formation on n-Si. In addition, several authors have investigated the photodissolution of n-type silicon in HF-containing electrolyte [7,18]. On n-type silicon, illumination is used to supply holes at the silicon surface and for pore growth. It depends on illumination intensity and wavelength. On p-type silicon, illumination may increase the conductivity of the substrate and of the porous layer [19]. Most of the present observations can be rationalized in terms of a photoinduced etching mechanism [20]. p-Type Si (p-Si) can also be modified by different illumination intensities [16]. The concentration of charge carriers and illumination intensity regulates the current density in the etching process. The current density on the silicon surface not only determines the rate of por-Si formation but also has a significant impact on its structure and physical characteristics. PhACE is characterized by low silicon etching rates, economic efficiency, and the ability to etch many wafers simultaneously. Only the uniform supply of reagents and light to the surface of silicon is required. There is no data on the etching rate and charge on the Si surface during PhACE of p-Si under visible light illumination in the literature. The originality of this paper is the formation of porous silicon and an investigation of the etching rate and total current on the Si surface during etching under front-side illumination and the dark. The paper provides information about the features of the porous silicon formation process in a potentiostatic regime.

2. Materials and Methods

Wafers of p-Si (100) with $\rho = 0.01$, 1, and 12 Ω·cm were used as the substrate. The silicon wafers were treated with piranha etch (a mixture of H_2SO_4 (98 wt.%) and H_2O_2 (30 wt.%) (1:2 by volume)) for 600 s at 130 °C. Then, the wafers were rinsed with deionized water and dried in an isopropyl alcohol vapor flow. The organic solvent vapor reduces the surface tension of the solution, and, due to this, water and small particles are removed from the surface of the substrate.

After that, the Si wafers were rinsed in an HF-H_2O solution (1:5 by volume) to remove native SiO_2. The wafers were then cut into pieces and used as samples. The etching was performed in a solution that contained HF (40 wt.%):H_2O_2 (30 wt.%):H_2O (2:1:10 by volume) at 25 °C for 20, 40, 60, and 80 min. The treatment was performed in the dark in an opaque Teflon beaker. The treatment was also performed with visible light using front-side illumination from a halogen lamp (JCDR 50 W) equal to 460 and 8000 lx. The distance between the lamp and the sample was 40 cm. The solution temperature was constant during illumination. The entire nonworking surface was protected by a chemically resistant varnish during etching. The entire working area was illuminated. Then, the samples were rinsed in an ethanol–water solution and dried in air. The sample surface morphology was investigated by scanning electron microscopy (SEM) using a JSM-6010PLUS/LA (JEOL company, Peabody, MA, USA) and Helios NanoLab 650 (Thermo Fisher Scientific, Waltham, MA, USA). AFM measurements were carried out in non-contact mode using a silicon cantilever at a resonance frequency of about 200 kHz under ambient conditions (microscope SOLVER-PRO, NT-MDT Ltd., Moscow, Russia) for roughness and pore walls analysis. AFM measurements were analyzed in Image Analysis. Gravimetric analysis was used for the porosity of the porous silicon calculation. The samples (five samples) were weighed before (m_1) and after (m_2) etching using analytical balance XP 205 (Mettler Toledo, Greifensee, Switzerland). The average values of the masses and the standard deviation were calculated:

$$\overline{m_n} = \frac{m_{n1} + m_{n2} + \ldots + m_{n5}}{5},$$

$$\Delta m = \sqrt{\frac{\sum_{i=1}^{5}(m_{ni} - \overline{m_n})^2}{4}}.$$

The sample area was measured ($S = a \times b$, a—length of the sample, b—width of the sample). It ranged from 0.5 to 1.5 cm^2. After etching, the porous silicon was dissolved in a 1 wt.% NaOH aqueous solution at room temperature until gas evolution stopped. After drying at room temperature, the samples were weighed (m_3). Porosity was calculated by the following equation:

$$P = \frac{m_1 - m_2}{m_1 - m_3} \times 100\%, \quad (1)$$

where m_1 and m_2 are the sample's masses before and after etching and m_3 is the sample mass after por-Si dissolution. Samples were etched in the same conditions because of the multi-sectional Teflon cell.

The short-circuit current in the galvanic cells was measured with a digital multimeter (UNI-T UT61C, Hong Kong, China). A Si sample connected with a Pt-electrode and set to a HF and $_{22}$-containing solution formed a galvanic cell. The value of the charge $Q'_{Excess\ Carrier}$ was determined by numerical integration of current versus time, $Q = \int_0^t Jdt$. The optical properties of the porous silicon formed by PhACE were measured using a UV–Vis–NIR spectrophotometer (Cary 5000, with an integrating sphere) in the 200–1300 nm wavelength range. Surface reflectance measurements were used to confirm that the porous silicon produced was useable for solar cells.

Gravimetric analysis was also used to define the number of carriers consumed over seconds during photo-assisted chemical etching. The investigations of excess current through silicon during PhACE with different illumination intensities were carried out using the short-circuit current. Short- circuit measurements were used to define current through the wafer during photo-assisted chemical etching. A mechanism for the formation of porous silicon without an external bias using the short-circuit measurement, *J(t)*, was established.

3. Results

It was found that porous silicon formation on p-Si is possible using front-side illumination without an external current source (electrochemistry) or metal on the surface of Si (MACE). The Si-HF-H_2O equilibrium diagram in [21] indicates band bending of p-Si in an HF solution, which leads to a hole-depleted layer formation on the Si surface. Therefore, a potential greater than the anodic silicon decomposition potential is required. Therefore, silicon anodic dissolution reactions are provided either by applying an external voltage, the deposition of a metal on the Si surface, or illumination. Photon absorption excites electrons to the conduction band, resulting in the formation of valence band holes. Silicon does not dissolve without an oxidizer (H_2O_2), because the role of H_2O_2 under photon excitation is to consume the excited conduction band electrons. The electrochemical potential of H_2O_2 is much more positive than the valence band of Si, and hole injection from H_2O_2 into the valence band is energetically possible [12,22]. This facilitates the transport of holes to the Si surface, which can then induce etching.

The mechanism of the photo-assisted etching of p- and n-Si with $\rho = 2\ \Omega\cdot$cm in HF was described in [23], with $\rho = 0.01$ and 10 $\Omega\cdot$cm [24]. Peroxide reduction is cathodic and silicon dissolution is an anodic process [25]:

$$H_2O_2 + 2H^+ + 2e^- \rightarrow 2H_2O, \tag{2}$$

$$Si + 6HF \rightarrow SiF_6^{2-} + 6H^+ + 4e^-. \tag{3}$$

Por-Si formation on the Si surface is possible due to the heterogeneous distribution of the electrochemical potential. There are some places on the surface of Si near which dissolution occurs at a higher rate. Doping level and resistivity influence the rate of pore formation. The lower the resistivity, the faster the etching rate. With the same modes of Si etching, silicon resistivity affects the amount of dissolved silicon. Additional illumination allows hole generation due to photon absorption. Porous silicon is the result of photo-assisted p-Si etching.

Figure 1 shows the time dependence of dissolved p-Si with 0.01, 1, and 12 $\Omega\cdot$cm for different illumination intensities ($\Delta m = m_1 - m_2$). The error was calculated as the standard deviation.

The effect of the resistivity, ρ, of doped silicon on the rate of its dissolution was noted. This effect is due to two factors. The inhomogeneous electrochemical potential distribution on the surface of the wafer is the first. The potential dispersion increases with impurity concentration in Si, so the pore density increases. Therefore, for p-Si with a lower resistivity, the dissolved mass will be higher than for p-Si with a high resistivity.

The silicon dissolution rate is directly proportional to the illumination intensity. The smaller value for the dissolved silicon mass at 0 lx is due to the low charge carrier generation rate at the Si surface in contact with the electrolyte. The maximum etching rate of Si was observed for p-Si with $\rho = 0.01$ $\Omega\cdot$cm under 8000 lx illumination. The rate of Si dissolution for $\rho = 0.01\ \Omega\cdot$cm is 3 and 11.8 µg/min, that for $\rho = 1\ \Omega\cdot$cm is 1.2 and 3.9 µg/min, and that for $\rho = 12\ \Omega\cdot$cm is 3 and 0.9 µg/min in the dark and under 8000 lx illumination, respectively. We approximated the dependence of the dissolved silicon mass change $\Delta m = m_1 - m_2$ on the etching time by a linear function. The angular coefficient of the linear approximation allowed the silicon etching rate for the solution to be calculated.

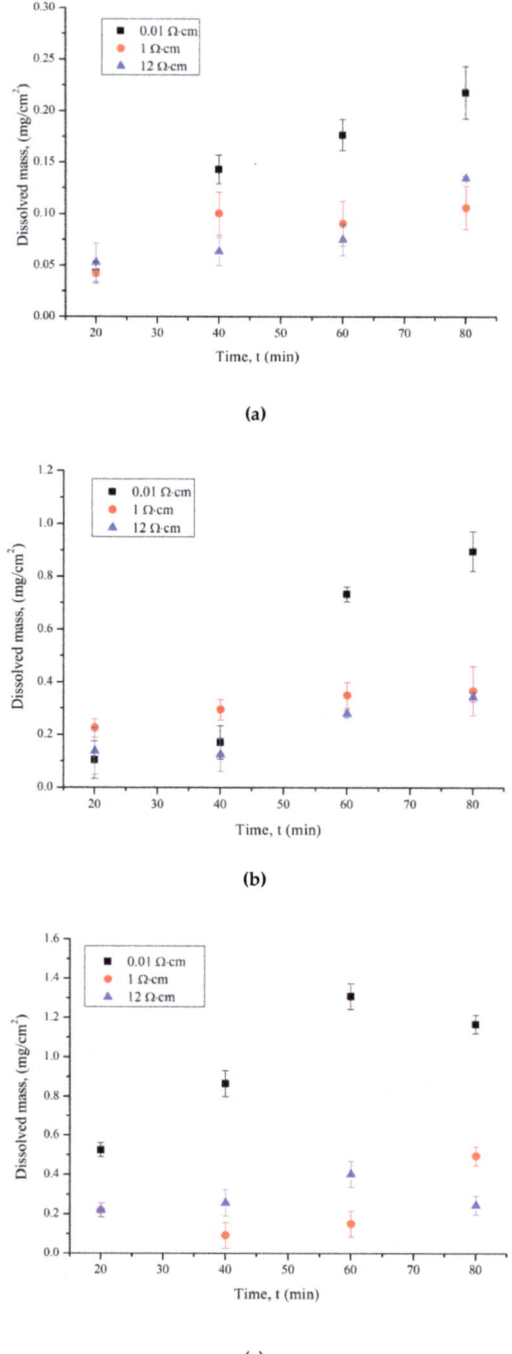

Figure 1. The effect of treatment duration on dissolved Si mass per surface area (**a**) in the dark, and under (**b**) 460 and (**c**) 8000 lx illumination.

Front-side illumination is necessary for porous silicon formation. The etching rate is determined by hole (h+) accumulation in the contact area of the HF electrolyte and Si atoms. The hole generation rate (g_x) in a semiconductor is given by the following equation [26]:

$$g_x = I_0 \cdot \alpha \cdot e^{-\alpha \cdot x}, \quad (4)$$

where I_0 is the intensity of light during etching, x is the depth of the light penetration, and α is the absorption coefficient.

According to Equation (4), the first factor that affects the hole generation rate is the light intensity during etching. The higher the illumination intensity, the higher the dissolution rate (Figure 1). Light absorptance (LA) is the second factor that affects the hole generation rate. Silicon wafers with ρ = 0.01, 1, and 12 Ω·cm before photo-assisted etching have a low absorption coefficient. The effect of short-term silicon etching in the solution containing HF and H_2O_2 is por-Si, which exhibits high LA properties [16,26] (Figure 2). LA depends on porous silicon thickness and pore diameters. Etching rate and resistivity influence these.

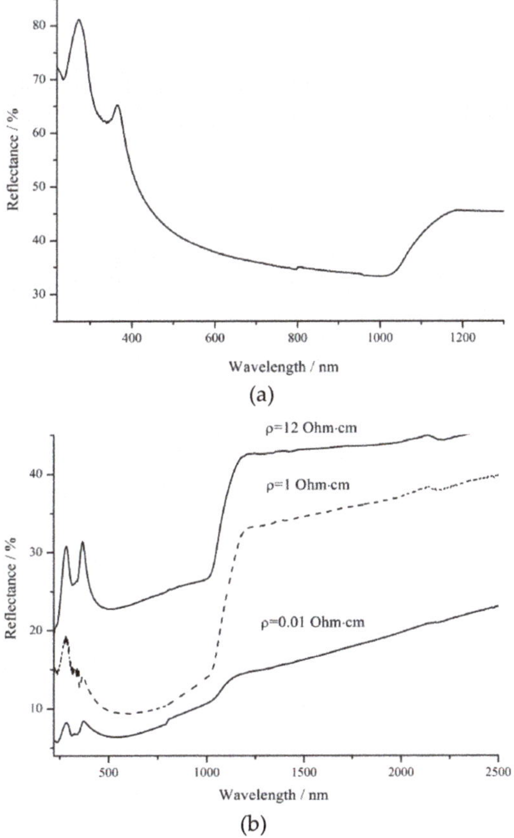

Figure 2. (**a**) Reflectance curve of the sample with ρ = 0.01 before treatment, and (**b**) reflectance curves of the samples with ρ = 0.01, 1, and 12 Ω·cm after 10 min of treatment.

Figure 2 shows the reflectance spectra of silicon before etching and porous silicon formed for 600 s under 8000 lx illumination.

As can be seen from the reflectance spectra, even after 600 s of porous silicon formation, the optical reflectance at the air–silicon interface is significantly reduced as the porous layer thickness increases. The resistivity influences the porous silicon thickness (Figure 1). Thus, the sample formed on p-Si with $\rho = 0.01\ \Omega{\cdot}$cm under 8000 lx illumination has a minimum reflectance, because the pore density has maximum values in this case. Porous silicon thickness depends on treatment duration and illumination.

The dissolution rate of Si is a function of the current flowing through the circuit, which characterizes the total number of charge carriers that are injected into silicon and participate in its dissolution over seconds. The total current (TC), J, on the Si surface during etching in the dark and under illumination can be calculated as follows [27]:

$$J = \frac{\frac{(m_1 - m_2)}{S} \cdot n \cdot e}{t \cdot m_{Si}} = \frac{V \cdot n \cdot e}{m_{Si}}, \tag{5}$$

where m_1 is the mass before etching, m_2 is the mass after etching during t, $(m_1 - m_2)/S$ is result of gravimetric analysis, m_{Si} is mass of the Si atom, n is the number of holes required for dissolution of each silicon atom, which equals 2 [28], t is the etching duration, e is the elementary electric charge, and S is the sample area.

Figure 3 shows that the TC, J, calculated using Equation (5), occurs on the p-Si surface during PhACE under different illumination intensities. The current decreases with increasing treatment time as the rate of porous silicon formation slows down. There are two possible reasons. First, the porous silicon layer absorbs light. Light does not reach the bottom of the pores, which leads to dissolution of the pore walls but not growth of the porous layer thickness. The second reason is that with an increase in thickness of the porous layer, reagents take longer to reach the chemical reaction zone at the bottom of the pores, so the concentration of HF and H_2O_2 decreases, as does the rate of silicon dissolution. This fact explains the form of the *J(t)* curve.

TC increases with the front-side surface illumination intensity. Light is suitable for creating majority charge carriers (holes) in silicon, which are necessary for etching. Thus, the values of the anode J depend on charge transfer mechanisms controlled by the illumination level and the resistivity of the initial Si wafer.

A p-type silicon wafer connected to platinum in a solution containing HF and H_2O_2 gives a galvanic cell [29,30]. Pt is in the electrolyte close to the sample surface. The current flows through the electrolyte between the Pt cathode and the silicon anode. This current can be measured using the short-circuit current. The cell produces a steady-state current proportional to the silicon area (Figure 4).

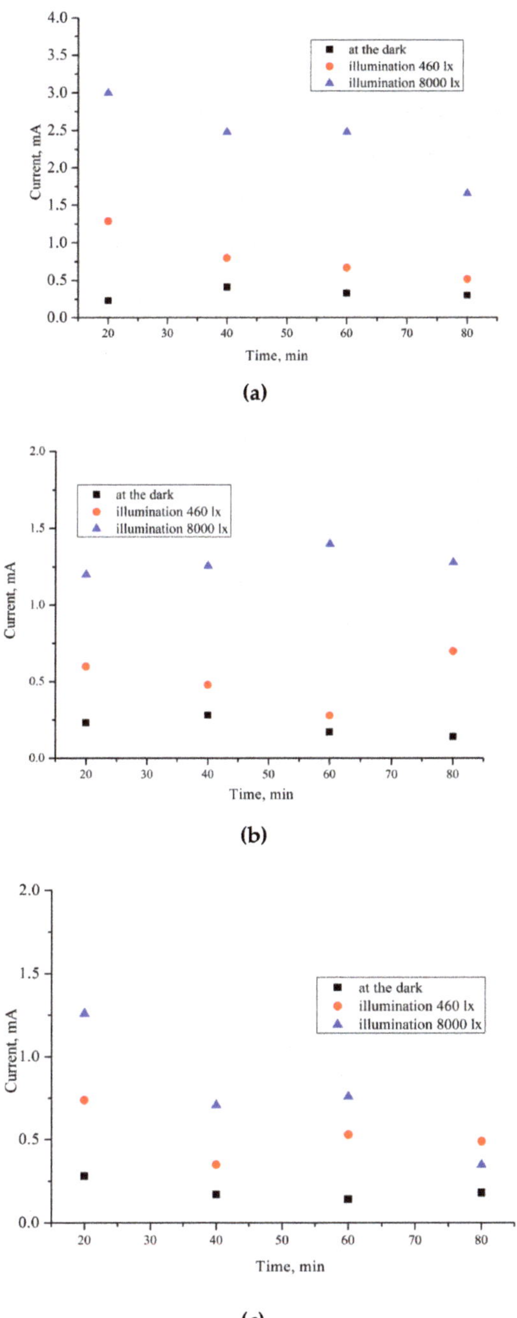

Figure 3. Current density, J, on the Si surface equaling 1 cm^2 after etching silicon with: (**a**) $\rho = 0.01$, (**b**) 1, () and 12 Ω·cm.

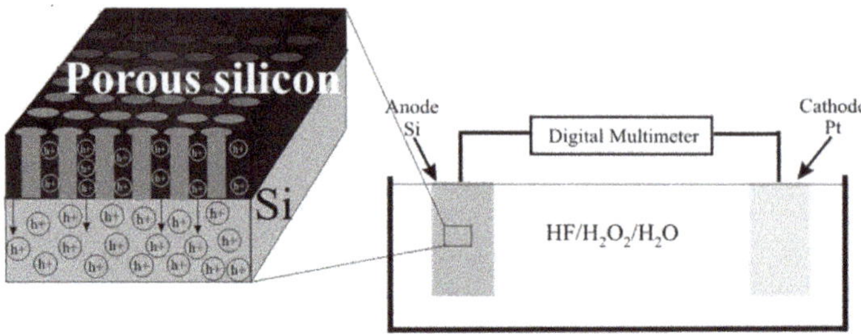

Figure 4. Schematic of the galvanic cell.

When the p-type silicon sample is connected to platinum in the HF/H_2O_2 solution, short-circuit current is observed in the dark and under illumination. Hydrogen peroxide is reduced during the etching. The holes injected into the valence band of silicon participate in the dissolution of Si and the formation of a porous layer. The transport of holes through porous silicon is difficult due to the high resistivity of the porous layers [31]. The excess holes that are not involved in the dissolution of Si diffuse into the semiconductor. The current of nonequilibrium charge carriers and the ion current can be measured by a galvanic cell (Figures 5 and 6). Figures 5 and 6 show the current measured after the Si and Pt are short-circuited in the dark and under illumination, respectively, for a 1 cm^2 sample.

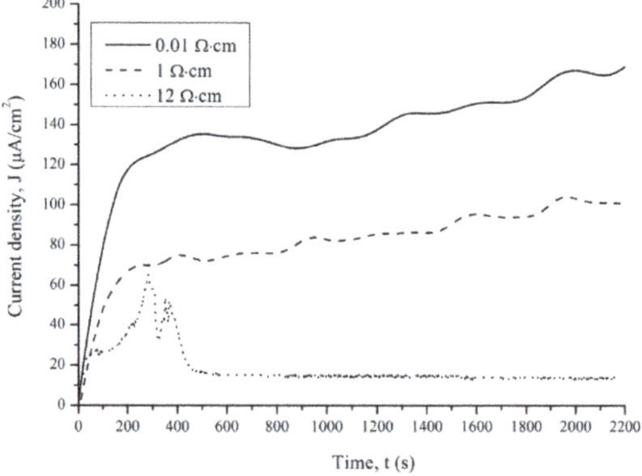

Figure 5. Current density measured in the dark.

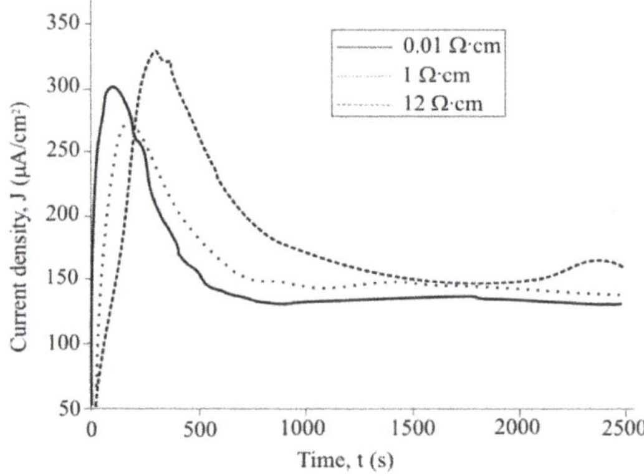

Figure 6. Current density measured under 460 lx illumination.

4. Discussion

The $J(t)$ curve measured under 460 lx illumination has a typical shape for the potentiostatic regime. The short-circuit current occurring under 460 lx illumination first rises sharply and then decreases significantly over 180–360 s. After 360 s, the current density is reduced from the maximum value to a stable value.

The $J(t)$ curves characterize the etching mechanism. Three characteristic regions can be identified:

1. Current increasing,
2. slowing growth rate and subsequent decrease in current,
3. constant current.

$J(t)$ reflects a change in the area (S) of the electrochemical reaction front. Changes in J with time (Figure 7) are related to the evolution of the Si morphology during pore nucleation. Pore formation takes place after the immersion of silicon into the solution containing HF and H_2O_2. The pore area depends on the duration of treatment. The increase in the specific surface area leads to growth of the current density—the first region. The current growth will continue until a porous layer of critical thickness is formed on the Si surface. In this case, access of the solution to the surface of monocrystalline silicon becomes limited. The transport of holes through porous silicon is difficult due to the high resistivity of the por-Si.

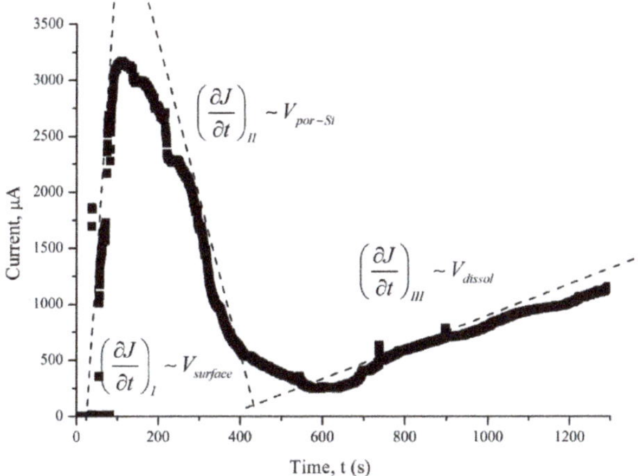

Figure 7. Current density measured under 8000 lx illumination.

With an increase in the thickness of the porous layer, the concentration of excess holes that are not involved in the dissolution of Si becomes smaller, which is reflected in the value of the slope angle of region II, which characterizes the growth rate of the por-Si layer. A further effect of the solution on the surface leads to the dissolution of the porous layer (the beginning of region III), and the illumination of the sample will play an important role. Dissolution of por-Si will reduce the thickness of the porous layer to less than the critical value, which will increase the concentration of holes in Si and, consequently, the current in region III. The slope angle of the section will characterize the dissolution rate of the porous V_{dissol} layer. That is why the inclination angle of region III $J(t)$ measured under 460 lx (Figure 6) is less than that under 8000 lx illumination (Figure 7).

The slope of the I segment ($\partial J/\partial t$) characterizes the porous layer nucleation rate, which will be higher for $\rho = 0.01\ \Omega{\cdot}\text{cm}$ in the case of different values of ρ. This result is consistent with the experimental data in Figure 6. The higher the speed of the $V_{surface}$, the greater the area of silicon dissolution, so a larger number of charge carriers will inject into silicon and reach the non-working side of the plate.

When comparing the results obtained from calculations using Equation (5) and measurements in a galvanic cell, a significant difference in currents was found. The value of $Q'_{Excess\ Carrier}$ was determined by numerical integration of the dependence of the current on time (Figure 6). The $Q'_{Excess\ Carrier}$ value was low compared to Q_{total} as it characterizes the current of excess charge carriers that have diffused into the substrate. Table 1 presents the charge values Q_{total}, $Q'_{Excess\ Carrier}$, and ΔQ for samples with different treatment parameters.

Table 1. Charge values for porous silicon samples.

Value	Illumination, lx	$\rho = 0.01\ \Omega{\cdot}cm$	$\rho = 1\ \Omega{\cdot}cm$	$\rho = 12\ \Omega{\cdot}cm$
Q_{total} (Equation (4))	0	1.535 ± 0.953	0.688 ± 0.282	0.437 ± 0.024
$Q'_{Excess\ Carrier}$ $(J(t))$		0.27	0.16	0.039
ΔQ (Equation (5))		1.265	0.528	0.398
Q_{total}	460	2.02 ± 1.3	1.17 ± 0.043	0.86 ± 0.437
$Q'_{Excess\ Carrier}$		0.348	0.37	0.496
ΔQ		1.674	0.8	0.364
Q_{total}	8000	5.95 ± 4.475	2.95 ± 0.0062	1.77 ± 0.948
$Q'_{Excess\ Carrier}$		5.28	2.05	1.64
ΔQ		0.67	0.9	0.13

The charge passing through the substrate is a term in Equation (6) and depends on the concentration of charge carriers injected into silicon Q_{total}.

$$Q'_{Excess\ Carrier} = Q_{total} - Q. \tag{6}$$

Q_{total}, calculated by Equation (5): $Q = J{\cdot}t$, characterizes the value of all charge carriers involved in the etching process. The increase in charge carrier concentration on the surface of the hole caused by injection leads to the appearance of a diffusion electron flow directed along the x-axis perpendicular to the semiconductor surface, with the result that the carrier concentration increases not only on the surface but also in the depth of the semiconductor. In this case, injected carriers go deeper into the semiconductor at different distances, where they are recombined. The contribution of the morphology of the porous layer is significant in the value of $Q'_{Excess\ Carrier}$ as some of the charge carriers remain in the porous layer ΔQ.

Having obtained the values of ΔQ, it is possible to determine the effect of light intensity and resistivity to estimate the change in the thickness of the porous layer. The decrease in ΔQ with an increase in the resistivity during etching in the dark and 460 lx illumination is associated with a decrease in the thickness of porous silicon. This is confirmed by the results of gravimetric analysis. The high value of $Q'_{Excess\ Carrier}$ and low ΔQ samples etched at 8000 lx, in relation to the above, are associated with an intensive dissolution of the porous layer, as well as an increase in the diameter of the pores in the latter. The increase in the diameter of pores contributes to the improvement in the access of the reagents to the silicon surface, which increases the concentration of charge carriers injected into the semiconductor.

As can be seen from the SEM images in Figure 8, the thickness of porous silicon formed in the dark reaches values not exceeding 300 nm, at 460 lx—not more than 1 micron, and at 8000 lx—the dissolution of the porous layer is clearly observed in both thickness and in the pore walls thickness. Illumination of the reaction zone is necessary for the thickness and porosity of porous silicon growth. However, long-term 8000 lx illumination contributes to the pore diameter growth, which leads to a reflectance increase.

Figure 8. Cross-section of samples formed in the (**a**) dark, and (**b**) 460 and (**c**) 8000 lx.

Figure 9 shows SEM (a–c) and AFM images (d–f) of the samples with $\rho = 0.01$ Ω·cm after different etching regimes. The pore diameter increases after etching under high 8000 lx illumination.

In this case, the intense generation of holes occurs in the pore walls, which contributes to wall dissolution. The walls dissolve, and the pores gradually unite with each other. Such samples differ from black porous silicon (3600 s, 460 lx); they are brown. «Black» and «brown» porous silicon are characterized by the color of the samples. The roughness and pore wall size of the black and brown silicon were measured by AFM using the non-contact mode. AFM measurements were analyzed in Image Analysis, resulting in the roughnesses presented in Figure 9d–f. The thickness of the walls of black and brown silicon is shown in Figure 9g.

As can be seen from Figure 9 and Table 2, the pore sizes increase with treatment duration from 10–100 to 25–125 nm for 10 and 60 min, respectively. The illumination 8000 lx during the 60 min etching of silicon leads to the formation of a porous layer with 90–440 nm pore diameter. Thus, illumination intensity significantly influences the pore size. The illumination intensity and treatment duration influence the pore wall thickness. This is confirmed by AFM measurement results. Black porous silicon has a wall thickness of 150–175 nm; brown porous silicon has a 350–375 nm wall thickness, because of wall dissolution and porous under-layer formation.

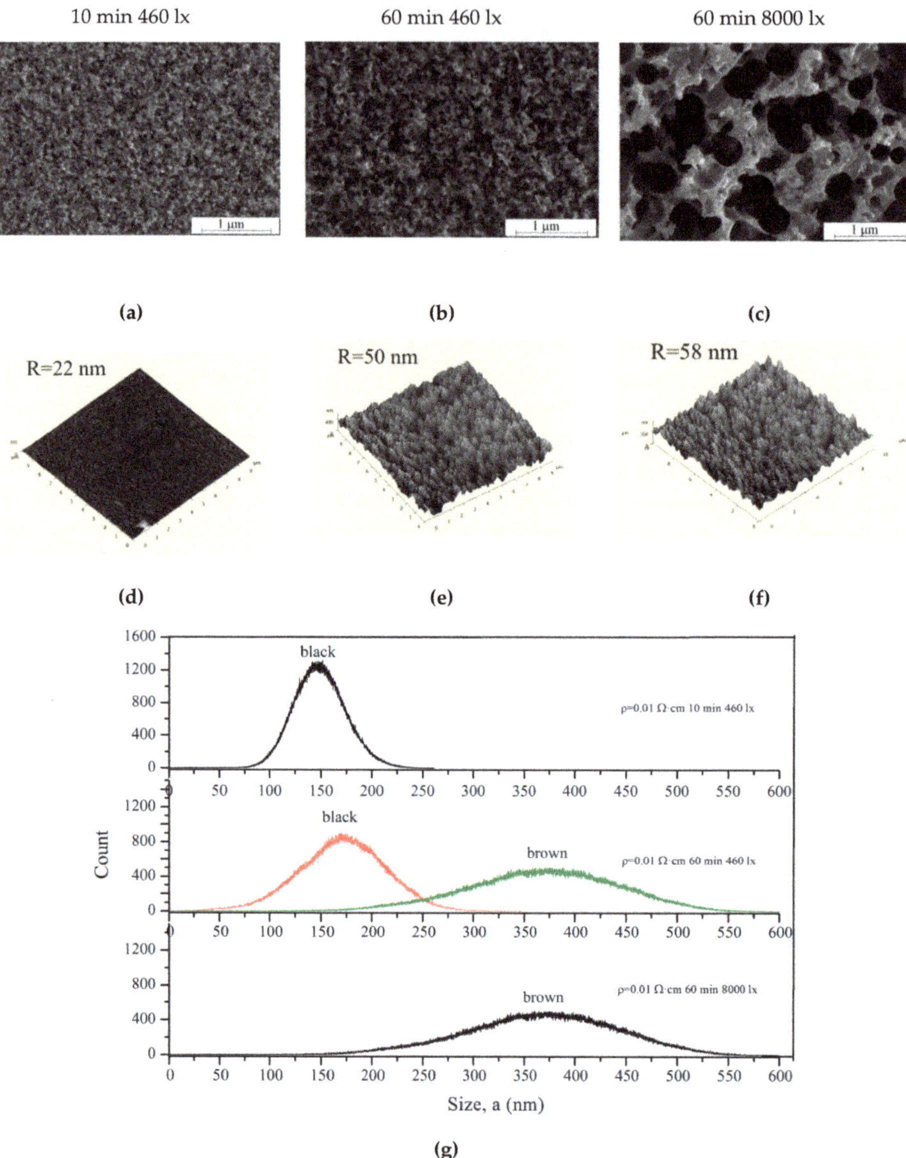

Figure 9. Morphology evolution of porous silicon by photo-assisted etching inspected after (**a**, **d**) 10 and (**b**, **c**, **e**, **f**) 60 min under (**a**, **b**, **d**, **e**) 460 and (**c**, **f**) 8000 lx illumination. (**a**–**c**) SEM images, (**d**–**f**) AFM images, and the (**g**) wall thickness of black and brown silicon.

Table 2. The roughness and the pore diameters of black and brown silicon.

№	Treatment duration, min	Illumination, lx	P, % (gravimetric analysis)	Roughness, nm (AFM)	d pores, nm (SEM)
1	10	460	14	22	10–100
2	60	460	60	50	25–125
3	60	8000	68	58	90–440

The increase in roughness and porosity contributes to the deterioration of the optical properties of porous silicon for solar cell applications (Figure 10). In the visible (VIS) region, the refractive index decreases with porosity, in contrast to the increasing edge absorption coefficient with an increase in porosity [32]. The porous silicon layer further increases the efficiency and absorbs more light energy in c-Si solar cells [33].

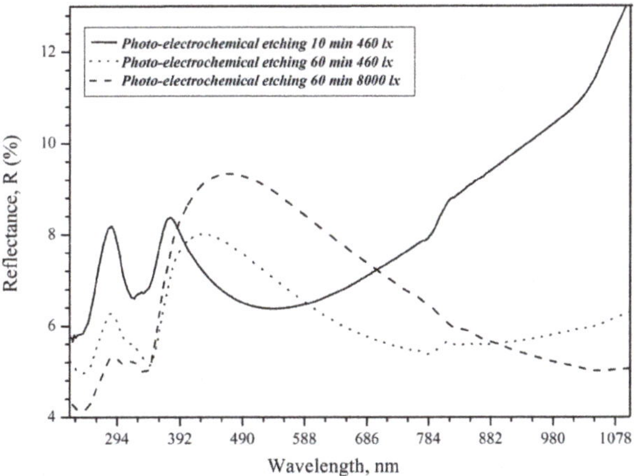

Figure 10. Reflectance curves of the samples with $\rho = 0.01\ \Omega\cdot$cm after 10 and 60 min treatment under 460 and 8000 lx illumination.

Reflection peaks (8.4%) 280 and 370 nm are associated with direct interband transitions in c-Si in the case of 73 nm thick porous layers [34]. At these wavelengths, the reflectance decreases to 6.4 and 5.4% as the thickness increases due to rising treatment duration. The effect of changing the thickness is now critical for shorter wavelengths [35]. The reflection peak at 370 nm shifts along the x-axis to 425 and 480 nm for samples after 3600 s etching under 460 and 8000 lx illumination, respectively. The roughness of the surface can influence the shift in the position of the maximum. Hence, it can be concluded that the reflectance increase in the visible spectral range, as shown in Figure 8, is solely the result of the surface roughness induced by illumination, which is consistent with literature results [36]. As a result, an 8000 lx illumination of silicon during etching promotes an increase in the roughness and reflection of the porous silicon surface, which is then called brown.

Thus, a comprehensive study of the currents involved in the dissolution of silicon, short-circuit currents, and optical and morphological properties made it possible to identify the effect of illumination intensity and resistivity on the concentration of nonequilibrium charge carriers. The absence of an external source of current and a metal catalyst on the surface made it possible to establish a significant contribution to the illumination intensity and resistivity as separate parameters of the etching process. The morphology and optical properties of the porous layer were shown to have a significant effect on the short-circuit current and, accordingly, on the concentration of non-equilibrium charge carriers in the semiconductor wafer.

5. Conclusions

It has been shown in this paper that porous silicon of different thicknesses, pore diameters, and porosity can be effectively fabricated by the photo-assisted etching on a Si surface without external bias or metals. The morphology of porous layers can be modified by varying parameters such as etching duration, surface illumination intensity, and wafer resistivity. Charge carriers that occur in Si during etching under illumination and in the dark lead to black and brown silicon formation. The total current on the Si surface can reach 3 mA for $\rho = 0.01$ Ω·cm and 8000 lx illumination. The total current value is comparable to the current during electrochemical etching. Por-Si can be used to enhance optical absorption in the UV, visible, and near-IR spectra. The results obtained in this work will allow the development of technology to increase the efficiency of the solar cells at the p–n junction due to the formation of black silicon on the Si surface. Control of the porous layer thickness with a minimum reflection coefficient will be carried out by measuring the current using the short-circuit current.

Author Contributions: Conceptualization, O.V.; methodology, O.V.; validation, O.V.; formal analysis, O.V.; investigation, P.L.; resources, O.V. and P.L.; writing—original draft preparation, O.V.; writing—review and editing, S.G. and O.V.; visualization, O.V.; supervision, S.G.; funding acquisition, O.V. All authors have read and agreed to the published version of the manuscript.

Funding: This investigation was supported by the Russian Science Foundation (project № 19-79-00205).

Acknowledgments: The authors are grateful to Alexey Romashkin from the National Research University of Electronic Technology (MIET), Moscow, Russian for AFM measurements; Alexander Dudin from the Institute of Nanotechnology of Microelectronics, Russian Academy of Sciences, Moscow, Russia for SEM measurements.

Conflicts of Interest: The authors declare that they have no conflicts of interest.

References

1. He, W.; Yurkevich, I.V.; Canham, L.T.; Loni, A.; Kaplan, A. Determination of excitation profile and dielectric function spatial nonuniformity in porous silicon by using WKB approach. *Opt. Express* **2014**, *22*, 27123–27135. [CrossRef] [PubMed]
2. Levitsky, I.A. Porous Silicon Structures as Optical Gas Sensors. *Sensors* **2015**, *15*, 19968–19991. [CrossRef] [PubMed]
3. Vashpanov, Y.; Jung, J.; Kwack, K. Photo-EMF Sensitivity of Porous Silicon Thin Layer–Crystalline Silicon Heterojunction to Ammonia Adsorption. *Sensors* **2011**, *11*, 1321–1327. [CrossRef] [PubMed]
4. Ikonen, T.; Nissinen, T.; Pohjalainen, E.; Sorsa, O.; Kallio, T.; Lehto, V.-P. Electrochemically anodized porous silicon: Towards simple and affordable anode material for Li-ion batteries. *Sci. Rep.* **2017**, *7*, 7880. [CrossRef]
5. Toor, F.; Oh, J.; Branz, H.M. Efficient nanostructured 'black' silicon solar cell by copper-catalyzed metal-assisted etching. *Prog. Photovolt. Res. Appl.* **2015**, *23*, 1375–1380. [CrossRef]
6. Barillaro, G.; Nannini, A.; Piotto, M. Electrochemical etching in HF solution for silicon. *Sens. Actuators A* **2002**, *102*, 195–201. [CrossRef]
7. Lehmann, V.; Föll, H. Formation Mechanism and Properties of Electrochemically Etched Trenches in n-Type Silicon. *J. Electrochem. Soc.* **1990**, *137*, 653–659. [CrossRef]
8. Gavrilov, S.A.; Karavanskii, V.A.; Sorokin, I.N. Effect of the electrolyte composition on properties of porous silicon layers. *Russ. J. Electrochem.* **1999**, *35*, 729–734.
9. Ashruf, C.M.A.; French, P.J.; Bressers, P.M.M.C.; Sarro, P.M.; Kelly, J.J. A new contactless electrochemical etch-stop based on a gold/silicon/TMAH galvanic cell. *Sens. Actuators A* **1998**, *66*, 284–291. [CrossRef]
10. Pyatilova, O.V.; Gavrilov, S.A.; Shilyaeva, Y.I.; Pavlov, A.A.; Shaman, Y.P.; Dudin, A.A. Influence of the Doping Type and Level on the Morphology of Porous Si Formed by Galvanic Etching. *Semiconductors* **2017**, *51*, 173–177. [CrossRef]
11. Kolasinski, K.W. The mechanism of galvanic/metal-assisted etching of silicon. *Nanoscale Res. Lett.* **2014**, *9*, 432. [CrossRef] [PubMed]
12. Huang, Z.; Geyer, N.; Werner, P.; Boor, J.; Gösele, U. Metal-Assisted Chemical Etching of Silicon: A Review. *Adv. Mater.* **2011**, *23*, 285. [CrossRef] [PubMed]
13. Li, X.; Bohn, P.W. Metal-assisted chemical etching in HF/H_2O_2 produces porous silicon. *Appl. Phys. Lett.* **2000**, *77*, 285–308. [CrossRef]

14. Zhang, M.L.; Peng, K.Q.; Fan, X.; Jie, J.S.; Zhang, R.Q.; Lee, S.T.; Wong, N.B. Preparation of large-area uniform silicon nanowires arrays through metal-assisted chemical etching. *J. Phys. Chem. C* **2008**, *112*, 4444–4450. [CrossRef]
15. Thomssen, M.; Berger, M.G.; Arens-Fischer, R.; Guck, O.; Krtiger, M.; Liath, H. Illumination-assisted formation of porous silicon. *Thin Solid Film.* **1996**, *276*, 21–24. [CrossRef]
16. Volovlikova, O.V.; Gavrilov, S.A.; Sysa, A.V.; Savitskiy, A.I.; Berezkina, A.Y. Ni-activated photo-electrochemical formation of por-Si in $HF/H_2O_2/H_2O$ Solution. In Proceedings of the 2017 IEEE Conference of Russian Young Researchers in Electrical and Electronic Engineering (EIConRus), St. Petersburg, Russia, 1–3 February 2017; pp. 1213–1216.
17. Koker, L.; Kolasinski, K.W. Laser-Assisted Formation of Porous Si in Diverse Fluoride Solutions: Reaction Kinetics and Mechanistic Implications. *J. Phys. Chem. B* **2001**, *105*, 3864–3871. [CrossRef]
18. Lévy-Clément, C.; Lagoubi, A.; Tomkiewicz, M. Morphology of Porous n-Type Silicon Obtained by Photoelectrochemical Etching. *J. Electrochem. Soc.* **1994**, *141*, 958–967. [CrossRef]
19. Cheah, K.W.; Choy, C.H. Wavelength dependence in photosynthesis of porous silicon dot. *Solid State Commun.* **1994**, *91*, 795–797. [CrossRef]
20. Chiboub, N.; Gabouze, N.; Chazalviel, J.-N.; Ozanam, F.; Moulay, S.; Manseri, A. Nanopore formation on low-doped p-type silicon under illumination. *Appl. Surf. Sci. Vol.* **2010**, *256*, 3826–3831. [CrossRef]
21. Gavrilov, S.A.; Belogorokhov, A.I.; Belogorokhova, L.I. A mechanism of oxygen-induced passivation of porous silicon in the HF: HCl: C2H5OH solutions. *Semiconductors* **2002**, *36*, 98–101. [CrossRef]
22. Wang, D.; Ji, R.; Du, S.; Albrecht, A.; Schaaf, P. Ordered arrays of nanoporous silicon nanopillars and silicon nanopillars with nanoporous shells. *Nanoscale Res. Lett.* **2013**, *8*, 42. [CrossRef] [PubMed]
23. Andersen, O.K.; Frello, T.; Veje, E. Photoinduced synthesis of porous silicon without anodization. *J. Appl. Phys.* **1995**, *78*, 6189–6192. [CrossRef]
24. Noguchi, N.; Suemune, I. Luminescent porous silicon synthesized by visible light irradiation. *Appl. Phys. Lett.* **1993**, *62*, 1429–1431. [CrossRef]
25. Qu, Y.; Liao, L.; Li, Y.; Zhang, H.; Huang, Y.; Duan, X. Electrically conductive and optically active porous silicon nanowires. *Nano Lett.* **2009**, *9*, 4539–4543. [CrossRef] [PubMed]
26. Narayanan, P. Photoelectrochemical etching of isolated, high aspect ratio microstructures in n-type silicon (100). Master's Thesis, University of Mumbai, Mumbai, India, 2007.
27. Saha, H.; Dutta, S.K.; Hossain, S.M.; Chakraborty, S.; Saha, A. Mechanism and control of formation of porous silicon on p-type Si. *Bull. Mater. Sci.* **1998**, *21*, 195–201. [CrossRef]
28. Turner, D.R. Electropolishing Silicon in Hydrofluoric Acid Solutions. *J. Electrochem. Soc.* **1958**, *105*, 402–408. [CrossRef]
29. Xia, X.H.; Ashruf, C.M.A.; French, P.J.; Kelly, J.J. Galvanic Cell Formation in Silicon/Metal Contacts: The Effect on Silicon Surface Morphology. *Chem. Mater.* **2000**, *12*, 1671–1678. [CrossRef]
30. Ashruf, C.M.A.; French, P.J.; Sarro, P.M.; Kazinczi, R.; Xia, X.H.; Kelly, J.J. Galvanic etching for sensor fabrication. *J. Micromech Microeng.* **2000**, *9*, 505–515. [CrossRef]
31. Bisi, O.; Ossicini, S.; Pavesi, L. Porous silicon: A quantum sponge structure for silicon based optoelectronics. *Surf. Sci. Rep.* **2000**, *38*, 1–126. [CrossRef]
32. Santinacci, L.; Gonçalves, A.-M.; Simon, N.; Etcheberry, A. Electrochemical and optical characterizations of anodic porous n-InP(1 0 0) layers. *Electrochim. Acta* **2010**, *56*, 878–888. [CrossRef]
33. Badawy, W.A.; El-Sherif, R.M.; Khalil, S.A. Porous silicon layers—Preparation, Characterization and Morphology. *Electrochim. Acta* **2010**, *55*, 8563–8569. [CrossRef]
34. Amirtharaj, P.M.; Seiler, D.G. *Optical Properties of Semiconductors*; Optical Society Of America: New York, NY, USA, 1995.
35. Huet, A.; Ramirez-Gutierrez, C.F.; Rodriguez-García, M.E. Simulation of Reflectance of Porous Thin-Layer Systems. *Int. J. Comput. Appl.* **2015**, *111*, 43–48. [CrossRef]
36. Steglich, M.; Käsebier, T.; Zilk, M.; Pertsch, T.; Kley, E.-B.; Tünnermann, A. The structural and optical properties of black silicon by inductively coupled plasma reactive ion etching. *J. Appl. Phys.* **2014**, *116*, 173503-1–173503-12. [CrossRef]

Article

Design and Fabrication of Flexible Naked-Eye 3D Display Film Element Based on Microstructure

Axiu Cao , Li Xue, Yingfei Pang, Liwei Liu, Hui Pang, Lifang Shi * and Qiling Deng

Institute of Optics and Electronics, Chinese Academy of Sciences, Chengdu 610209, China; longazure@163.com (A.C.); xueli2553@163.com (L.X.); yfpang7647@163.com (Y.P.); liuliweineko@163.com (L.L.); ph@ioe.ac.cn (H.P.); dengqiling@ioe.ac.cn (Q.D.)

* Correspondence: shilifang@ioe.ac.cn; Tel.: +86-028-8510-1178

Received: 19 November 2019; Accepted: 7 December 2019; Published: 9 December 2019

Abstract: The naked-eye three-dimensional (3D) display technology without wearing equipment is an inevitable future development trend. In this paper, the design and fabrication of a flexible naked-eye 3D display film element based on a microstructure have been proposed to achieve a high-resolution 3D display effect. The film element consists of two sets of key microstructures, namely, a microimage array (MIA) and microlens array (MLA). By establishing the basic structural model, the matching relationship between the two groups of microstructures has been studied. Based on 3D graphics software, a 3D object information acquisition model has been proposed to achieve a high-resolution MIA from different viewpoints, recording without crosstalk. In addition, lithography technology has been used to realize the fabrications of the MLA and MIA. Based on nanoimprint technology, a complete integration technology on a flexible film substrate has been formed. Finally, a flexible 3D display film element has been fabricated, which has a light weight and can be curled.

Keywords: naked-eye 3D; microstructure; flexible; film; fabrication

1. Introduction

With the development of science and technology, people are hoping to truly restore three-dimensional (3D) information of the object space. As a result, 3D display technology has emerged with the times and has become a research hotspot in the field of image displays [1–6].

As Wheatstone invented the first stereoscopic picture viewer in 1838, the technology of 3D displays has been developed for nearly 200 years [7]. In this development process, head-mounted 3D display technology is very mature in principle and technology [8–11], and there are a large number of commercial products. However, due to the need of wearing equipment, it is always inconvenient. Meanwhile, long-term use depending on the binocular parallax principle will lead to viewing fatigue, and viewers will feel dizzy. Therefore, it is an inevitable trend for the future development of naked-eye 3D display technology without wearing equipment.

Research groups have developed a variety of naked-eye 3D display technologies, including the grating 3D display technology of binocular parallax, holographic technology, and integrated imaging technology. Grating 3D display technology uses the principle of binocular parallax to produce a 3D sensation [12,13]. This technology has the advantages of low cost, simple structure, and easy implementation. However, because the left and right parallax images cannot be completely separated, the viewing area of the viewer is limited, and the 3D image can only be viewed in a relatively fixed position, lacking freedom. Therefore, it is only suitable for a single user and small range of motion.

By using the interference principle, holographic technology interferes the light wave reflected by the object with the reference light wave, and records it in the form of interference fringes to form a hologram [14]. When the hologram is illuminated by a coherent light source, the original light wave will be reproduced based on the diffraction principle, to form a realistic 3D image of the original object. However, the production of high-quality optical holograms requires a high-coherence laser, shockproof platform, and precise optical path setting. In addition, the ambient air flow will also affect the successful recording of holograms. Later, with the development of digital holography technology, using charge coupled devices (CCDs) instead of ordinary holographic recording materials to record holograms and using computer simulations instead of optical diffraction to realize object reproduction, the digitization of the whole process of hologram recording, storage, processing, and reproduction can be realized [15]. However, the resolution of digital holography using CCDs to record coherent light waves cannot be compared to that of holographic dry plates, so the resolution of holograms is relatively low, which seriously affects the image clarity.

The computer-generated hologram, which combines digital computing with modern optics, encoding the complex amplitude of the object light wave from a computer to computer-generated hologram (CGH), has unique advantages and good flexibility [16]. However, there is a large amount of data information and processing time for 3D objects, so it is necessary to select appropriate algorithms and coding methods to overcome [5,17]. In addition, the spatial light modulator plays a very important role in the experiment of CGH photoelectricity reproduction. It is necessary to overcome the influence of spatial light modulation on the quality of the reconstructed image [18,19]. At present, the CGH is still in the research stage of algorithm optimization to realize a 3D display of large scale and large field of view. It is still early to expect a commercial product based on holographic display devices.

Integrated imaging technology is also composed of two basic processes: Recording and reproduction. Unlike holographic technology, the recording and reproduction process do not require the participation of coherent light, which reduces the difficulty of the whole system. This technology was first proposed by Lippmann, a famous French physicist and Nobel Laureate in physics, in 1908 [20]. The microlens units in the microlens array (MLA) are used to record 3D object information from different perspectives to form a microimage array (MIA), and then the recorded MIA is placed on the focal plane of the MLA whose parameters are matched with a microlens in the recording process. The 3D image can be viewed by irradiating with scattered light according to the principle of optical reversibility and the fusion of the human brain. This technology can provide full parallax and a full-color image, without any special equipment, and the viewpoint provided is quasi-continuous. In a certain area, it can be viewed by many people, which has become an important development trend of naked-eye 3D displays.

The recording and reconstruction of 3D objects are formed through the interaction of tens of thousands or even hundreds of thousands of microimages. A camera array can be used to record the MIA. This method requires a large number of expensive and complex camera equipment, and the mechanical error between camera equipment will also affect the final imaging effect [21,22]. The optical recording method [23] uses an MLA to record the MIA, which is easy to be affected by surrounding environmental conditions such as brightness, sensitivity, and uniformity. The experimental operation is difficult, and the adjacent images are prone to crosstalk, resulting in a poor imaging effect of the final MIA. Then, the acquisition of the MIA based on 3D graphics software is proposed [24], which can realize the free crosstalk and high-resolution recording of the MIA. In addition, the display screen is generally used to display the recorded microimage array in 3D image reproduction. At present, the screen resolution of the mainstream high-definition display screen is 1920 × 1080, and the number of pixels per inch is 89, i.e., 89 ppi. Compared to the resolution limit of the human eye of 300 ppi, the resolution is still low. The granular distortion effect will be visible at a certain distance.

In this paper, based on the integrated imaging technology, a flexible naked-eye 3D display film element based on microstructures has been designed and fabricated, which can achieve the 3D imaging effect with a resolution higher than the human eye resolution limit of 300 ppi, and has the advantages of curling and light weight. The main arrangement of this paper is as follows: The second part describes the structure design and imaging principle of the 3D display film element. The third part presents the structural design of the MLA and acquisition of the MIA. The fourth part describes the preparation and integration of the microstructure, and the final part presents the summary of the whole paper.

2. Structural Design and Imaging Principle

The structure of the flexible naked-eye 3D display film element consists of three parts, namely, an MLA, MIA, and flexible substrate material, as shown in Figure 1a. The MIA is imaged by the MLA with different viewing angle information. The sub-images of each imaging channel are fused to form a 3D display effect. The human eye can watch the 3D image of the object in front of them, as shown in Figure 1b.

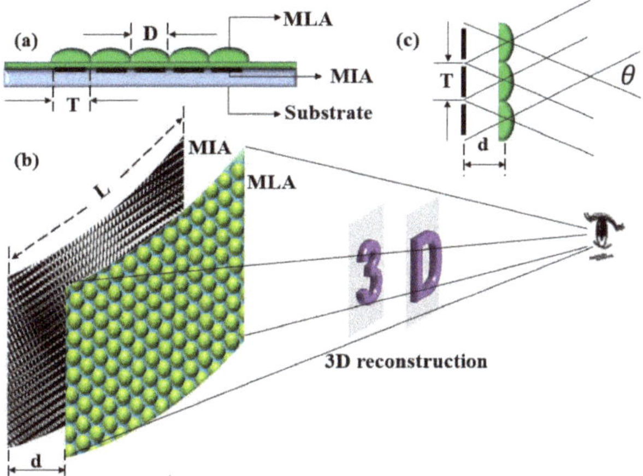

Figure 1. Flexible naked-eye 3D display film element: (**a**) structure composition; (**b**) imaging principle; (**c**) viewing angle.

The aperture (D) of the microlens is the same as the size (T) of the microimage unit, and the distance (d) between the MLA and MIA is the effective focal length (f) of the microlens. A single microlens can image the corresponding microimage independently, and several sub-images are fused to form a 3D effect. According to the theory of Gauss optics, when the distance between the MLA and MIA is the focal length of the microlens, the image distance is infinite, which means that light from any angle on the MIA is refracted by the MLA and then emitted as parallel light. Therefore, the number of pixels of the reconstructed 3D image is determined by the number of MLAs ($n \times n$). Finally, the imaging resolution (Re) of the element can be calculated according to Equation (1), where L is the size of the MIA. L can be obtained by multiplying the array number ($n \times n$) of microimages by the size of the microimage unit (T). The viewing angle of the film element (Figure 1c) can be obtained from Equation (2). As the focus display mode [25] is used in this paper, the 3D depth range (ΔZ) is determined by Equation (3), where P_I is the pixel size of the microimage.

$$\mathrm{Re} = \frac{n}{L}, \tag{1}$$

$$\theta = 2\arctan\left(\frac{T}{2d}\right), \tag{2}$$

$$\Delta Z = 2\frac{dD}{P_I}. \tag{3}$$

3. Structural Design of Microlens Array (MLA) and Acquisition of Microimage Array (MIA)

3.1. Structural Design of MLA

Due to the MLA being the key imaging element, the reasonable design of the parameters of the MLA, such as the aperture, focal length, and array number, is related to the integration of the whole element and the quality of the 3D image. For a miniaturized 3D film element, when the microlens is designed with a large aperture or a small number of MIAs, the resolution of images from all perspectives of the 3D image will be very low, which makes the viewing effect worse. In order to meet the requirement of the human eye resolution of 300 ppi, the structural parameters of the microlens are designed as follows: (1) the material of the microlens is a photosensitive adhesive (NOA61), and the refractive index is 1.56 in the visible light band; (2) the aperture (D) of the microlens is 80 μm; (3) the curvature radius of the microlens is 47.5 μm; (4) the focal length (f) of the microlens is 85 μm; (5) the sag height of the microlens is 22 μm; (6) the array number of the microlens is 250 × 250. Then, the imaging resolution is calculated as 317 ppi, which is higher than the human-eye resolution limit. The viewing angle (θ) is about 50°. At this time, the focal length of the microlens is very short, so the distance between the MLA and MIA is 85 μm. Thus, the whole element will reach the thin-film level.

3.2. Acquisition of MIA

The acquisition of tens of thousands or even hundreds of thousands of microimages has been carried out based on 3D graphics software. This method does not need complicated and expensive optical equipment, and can also avoid human and mechanical errors in the operation of optical equipment. With the use of computer memory, computer generation technology can generate microimages quickly, accurately, and in large quantities.

Based on 3ds MAX software (3ds MAX 2009, San Rafael, CA, USA), a 3D scene was established, as shown in Figure 2. The scene contained the Chinese characters "光电所" and letters "IOE." The central 3D coordinates of "光", "电", and "所" were (15.59 mm, −6.8 mm, 9.5 mm), (30.79 mm, 0, 9.5 mm), and (37.88 mm, 6.8 mm, 9.5 mm), respectively. The central 3D coordinates of "I", "O", and "E" were (37.35 mm, −6 mm, 13 mm), (23.43 mm, 0, 13 mm), and (12.48 mm, 6 mm, 13 mm), respectively. The virtual dynamic camera array was established to simulate the image acquisition process of the whole MLA, and the acquisition of the microimage was carried out for different 3D information from far to near. In the image acquisition, the 3D coordinates of the start point of the camera were A (0, −12.5 mm, 0), and the 3D coordinates of the end point were B (0, 7.42 mm, 19.92 mm). The field of view of the camera was 5° and the moving interval was 80 μm, which was matched with the structural parameters of the MLA. Finally, 250 × 250 microimages were acquired. The pixel number of the microimage was 40 × 40, and the pixel size of the microimage (P_I) was 2 μm. Therefore, the 3D depth range could be calculated as 6.8 mm.

In the process of MIA acquisition, the camera captured images of the 3D scene from different perspectives, as shown in Figure 3, to record the information at different perspectives. In addition, the obtained microimages are shown in the box in the upper right corner of the corresponding perspective. It can be seen that the microimages captured by the virtual camera have a very high image resolution and perfect image quality.

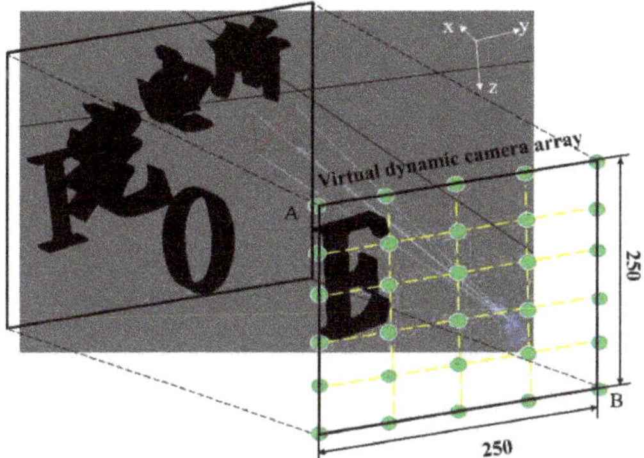

Figure 2. 250 × 250 microimages acquired by dynamic camera.

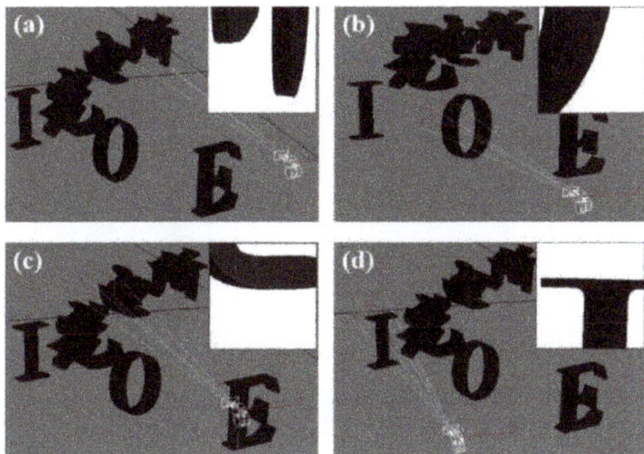

Figure 3. Imaging from different perspectives: (a) imaging from one prespective of "所", (b) imaging from one prespective of "O", (c) imaging from one prespective of "电", and (d) imaging from one prespective of "I".

Furthermore, 250 × 250 microimages captured by the camera were encoded and fused according to the arrangement of the MLA. The MIA was generated by computer processing, as shown in Figure 4. From the enlarged images of different regions, we can see that each microimage is different with information of different perspectives from the 3D scene.

Figure 4. Microimage array (MIA) with different enlarged images of different regions.

4. Preparation and Integration

There are two key microstructures in the 3D display element, which are the MLA and MIA. The MLA was prepared by photolithography and the hot melting method, and the preparation results are shown in Figure 5, which shows the prototype of the MLA (Figure 5a), micromagnifier of the microlens (Figure 5b), and surface profile of the microlens (Figure 5c). The sag height of the microlens was 21.97 μm, which is consistent with the design result. The MIA was also prepared by photolithography. The pattern was prepared on the substrate material by a series of processes such as exposure, development, and etching. The MIA and enlarged images of different regions of the MIA are shown in Figure 6. The substrate material was a polyethylene terephthalate (PET) film, which has characteristics of high toughness, smooth surface, and good light transmittance. The pattern was made by lithography with a high resolution of minimum linewidth of 2 μm, which is equal to the pixel size of the microimage.

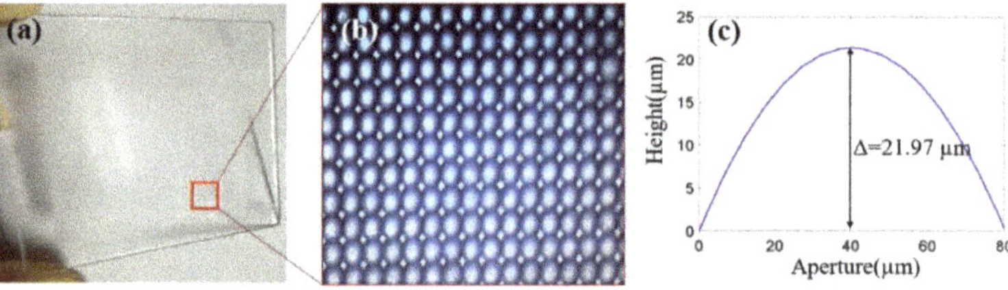

Figure 5. Preparation results of microlens: (**a**) prototype of microlens array (MLA); (**b**) micromagnifier of microlens; (**c**) surface profile of microlens.

Figure 6. Preparation results of (**a**) MIA in different areas: (**b**) few microimages of "I", (**c**) few microimages of "O", and (**d**) few microimages of "E".

Subsequently, integration of the two key microstructures needs to be carried out. During the integration, the MLA and MIA need to be aligned one by one to realize 3D image reconstruction. In the experiment, nano-imprinting alignment technology was used to achieve alignment integration of the two microstructures, as shown in Figure 7.

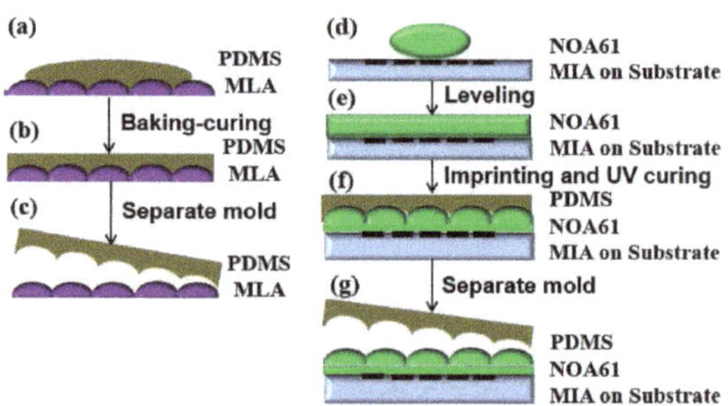

Figure 7. Integration of thin film element: (**a**) pouring of polydimethylsiloxane (PDMS) materials; (**b**) baking and curing; (**c**) generation of imprinting mold; (**d**) photosensitive adhesive (NOA61) dropped on the prepared MIA; (**e**) leveling; (**f**) imprinting and UV curing; (**g**) mold peeled off.

First, the imprinting mold with the structural information of the MLA should be prepared. The imprinting mold is composed of polydimethylsiloxane (PDMS). The free energy of the interface of the PDMS mold is low and has chemical inertness. Therefore, the mold is easy to be separated when the mold is used to carry out the integration. During the mold preparation, PDMS stroma and curing agent were poured into a clean beaker at a volume ratio of 10:1, continuously stirring with a glass rod. A large number of bubbles were generated in the PDMS prepolymer until the bubbles disappeared gradually. In addition, the PDMS prepolymer was poured on the prepared MLA, as shown in Figure 7a. Then, the substrate was placed on the coater at a speed of 250 rpm with a time of 20 s, shaking off the excess PDMS. Subsequently, it was placed in a vacuum oven until all the bubbles disappeared for curing, as shown in Figure 7b. The baking temperature and time were set as 65 °C and 4 h, respectively. Finally, the PDMS imprinting mold with negative structural information of the MLA on the surface could be peeled off from the MLA, as shown in Figure 7c.

Secondly, the structural information of the MLA should be imprinted on the surface of the MIA by using the imprinting mold to realize the integration of the MLA and MIA. The process flow was as follows:

First, the photosensitive adhesive (NOA61) was dropped on the prepared MIA, as shown in Figure 7d. After it was leveled (Figure 7e), the imprinting mold was used to imprint photosensitive adhesive. During imprinting, the high-precision alignment device was used to align the microlenses with the microimages one by one. On the basis of alignment, the photosensitive adhesive was irradiated by ultraviolet light with a wavelength of 365 nm until it was cured, as shown in Figure 7f. Finally, the PDMS mold was peeled off (Figure 7g) to obtain an integrated 3D display film element, as shown in Figure 8.

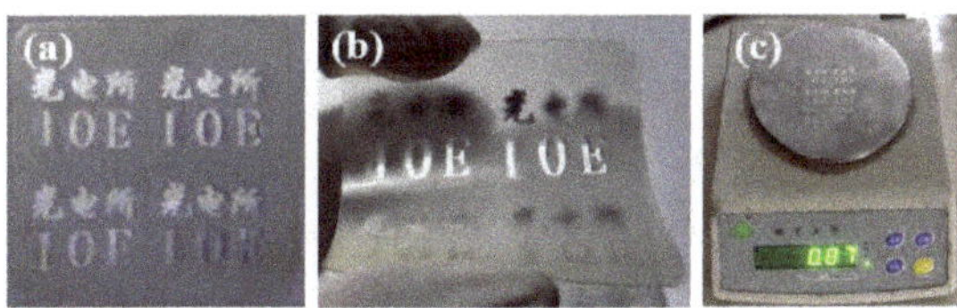

Figure 8. Integration of thin film element: (**a**) planar display; (**b**) curved display; (**c**) weight.

Figure 8a shows the 3D effect of the planar display, and the 3D display effect can be seen from various angles. Figure 8b shows the 3D display effect after bending of the element. Meanwhile, the weight of the 3D display element on the flexible substrate has been characterized, which is less than 1 g, as shown in Figure 8c, reaching the lightweight level. Figure 9 shows the 3D effects from different viewing angles.

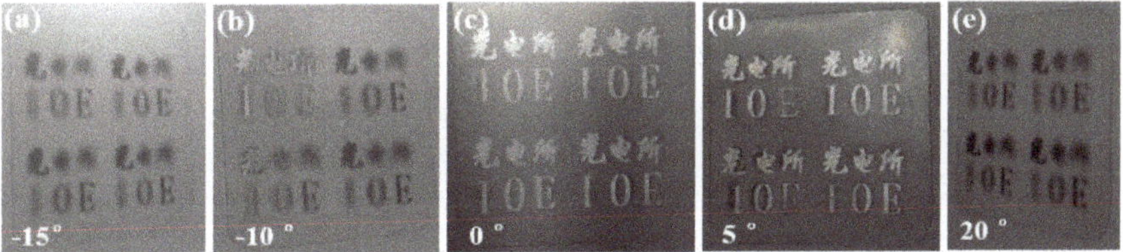

Figure 9. 3D effects from different viewing angles: (**a**) −15°; (**b**) −10°; (**c**) 0°; (**d**) 5°; (**e**) 20°.

5. Conclusions

In this paper, we propose to design and fabricate a 3D display film element based on microfabrication. The imaging resolution is higher than that of the human eye at 300 ppi. At the same time, the element has the characteristics of miniaturization and light weight, which can be applied to product packaging, handicrafts, anti-counterfeiting, and other industries. Using the 3D display effect to replace the original two-dimensional image display has a certain market application prospect, and also lays the technical foundation for wearable display equipment.

Author Contributions: Conceptualization, A.C.; Formal analysis, A.C. and L.X.; Funding acquisition, Q.D.; Investigation, L.X.; Software, Y.P.; Validation, H.P.; Visualization, L.L.; Writing—original draft, A.C.; Writing—review and editing, L.L. and L.S.

Funding: This research was supported by the National Natural Science Foundation of China (Nos. 61605211, 61905251); The Instrument Development of Chinese Academy of Sciences (No. YJKYYQ20180008); The National R&D Program of China (No. 2017YFC0804900); Sichuan Science and Technology Program (No. 2019YJ0014); Youth Innovation Promotion Association, CAS and CAS "Light of West China" Program. The authors thank their colleagues for their discussions and suggestions to this research.

Conflicts of Interest: The authors declare no conflict of interest.

References

1. Krebs, P.; Liang, H.; Fan, H.; Zhang, A. Homogeneous free-form directional backlight for 3D display. *Opt. Commun.* **2017**, *397*, 112–117. [CrossRef]
2. Yusuke, S.; Kazuo, S.; Takahiro, K.; Makoto, K.; Daisuke, B.; Toyohiko, Y. Super-wide viewing-zone holographic 3D display using a convex parabolic mirror. *Sci. Rep.* **2018**, *8*, 11333.
3. Xu, H.; Jiang, G.; Yu, M.; Luo, T.; Peng, Z.; Shao, F. 3D visual discomfort predictor based on subjective perceived-constraint sparse representation in 3D display system. *Future Gener. Comp. Syst.* **2018**, *83*, 85–94. [CrossRef]
4. Wenqiang, W.; Wen, Q.; Wenbin, H.; Ming, Z.; Yan, Y.; Xiangyu, C. Multiview holographic 3D dynamic display by combining a nano-grating patterned phase plate and LCD. *Opt. Express* **2017**, *25*, 1114–1122.
5. Zhen, Z.; Siqing, C.; Huadong, Z.; Zhenxiang, Z.; Hongyue, G.; Yingjie, Y. Full-color holographic 3D display using slice-based fractional Fourier transform combined with free-space Fresnel diffraction. *Appl. Opt.* **2017**, *56*, 5668–5675.
6. Zaperty, W.; Kozacki, T.; Kujawińska, M. Multi-SLM color holographic 3D display based on RGB spatial filter. *J. Disp. Technol.* **2017**, *12*, 1724–1731. [CrossRef]
7. Wheatstone, C. Contributions to the Physiology of Vision. Part the First. On Some Remarkable, and Hitherto Unobserved, Phenomena of Binocular Vision. *Philos. T. R. Soc. A* **1838**, *128*, 371–394.
8. Steinicke, F.; Bruder, G.; Hinrichs, K.; Lappe, M.; Kuhl, S.; Willemsen, P. Judgment of natural perspective projections in head-mounted display environments. *IEEE T. Vis. Comput. Gr.* **2009**, *17*, 888–899. [CrossRef]
9. Reike, I.; Riemann, B. Three-dimensional multi-view large projection system. *Proc. SPIE* **2005**, *5664*, 147.
10. Jung, S.M.; Park, J.U.; Lee, S.C.; Kim, W.S.; Chung, I.J. 25.4L: Late-News Paper: A Novel Polarizer Glasses-Type 3D Displays with an Active Retarder. *SID Int. Symp. Dig. Tech. Pap.* **2009**, *40*, 348–351. [CrossRef]
11. Keller, K.; State, A.; Fuchs, H. Head Mounted Displays for Medical Use. *J. Disp. Technol.* **2008**, *4*, 468–472. [CrossRef]
12. Park, D.; Kim, K.; Lee, C.; Son, J.; Lee, Y. Lenticular stereoscopic imaging and displaying techniques with nospecial glasses. *IEEE Image Process.* **1995**, *537599*, 137–139.
13. Son, J.Y. Viewing zones in three-dimensional imaging systems based on lenticular, parallax-barrier, and microlens-array plates. *Appl. Opt.* **2004**, *43*, 4985–4992. [CrossRef]
14. Guo, X.; Zhu, J.; Xia, C.; Li, J.; Chen, L. Characterization of a real-time high-sensitivity photopolymer for holographic display and holographic interferometry. *Proc. SPIE* **2005**, *5636*, 528–537.
15. Goodman, J.W.; Lawrence, R.W. Digital image formation from electronically detected holograms. *Appl. Phys. Lett.* **1967**, *11*, 77–79. [CrossRef]
16. Ruidan, K.; Juan, L.; Gaolei, X.; Xin, L.; Dapu, P.; Yongtian, W. Curved multiplexing computer-generated hologram for 3D holographic display. *Opt. Express* **2019**, *27*, 14369–14380.
17. Xue, G.; Liu, J.; Li, X.; Jia, J.; Zhang, Z.; Hu, B. Multiplexing encoding method for full-color dynamic 3D holographic display. *Opt. Express* **2014**, *22*, 18473. [CrossRef]
18. Makey, G.; Yavuz, Ö.; Kesim, D.K.; Turnalı, A.; Elahi, P.; Ilday, S.; Tokel, O.; Ilday F, Ö. Breaking crosstalk limits to dynamic holography using orthogonality of high-dimensional random vectors. *Nat. Photonics* **2019**, *13*, 251–256. [CrossRef]

19. Park, J.; Lee, K.R.; Park, Y.K. Ultrathin wide-angle large-area digital 3D holographic display using a non-periodic photon sieve. *Nat. Commun.* **2019**, *10*, 1304. [CrossRef]
20. Lippmann, G. Epreuves reversibles donnant la sensation durelief. *J. Phys.* **1908**, *7*, 821–825.
21. Papageorgas, P.G.; Athineos, S.S.; Sgouros, N.P.; Theofanous, N.G. 3D capturing devices based on the principles of Integral Photography. 2006. Available online: http://citeseerx.ist.psu.edu/viewdoc/summary?doi=10.1.1.160.3029 (accessed on 2 June 2019).
22. Deng, H.; Wang, Q.H.; Li, D.H.; Wang, F.N. Realization of Undistorted and Orthoscopic Integral Imaging Without Black Zone in Real and Virtual Fields. *J. Disp. Technol.* **2011**, *7*, 255–258. [CrossRef]
23. Ives, H.E. Optical Properties of a Lippmann Lenticulated Sheet. *J. Opt. Soc. Am. A* **1931**, *21*, 171–176. [CrossRef]
24. Jiao, T.T.; Wang, Q.H.; Li, D.H.; Zhou, L.; Wang, F.N. Computer-Generated Integral Imaging Based on 3DS MAX. *Chin. J. Liq. Cryst. Displays* **2008**, *23*, 621–624.
25. Hong, J.; Kim, Y.; Choi, H.J.; Hahn, J.; Park, J.H.; Kim, H. Three-dimensional display technologies of recent interest: Principles, status, and issues. *Appl. Opt.* **2011**, *50*, 87–115. [CrossRef]

Review

Emerging Designs of Electronic Devices in Biomedicine

Maria Laura Coluccio [1], Salvatore A. Pullano [2], Marco Flavio Michele Vismara [2], Nicola Coppedè [3], Gerardo Perozziello [1], Patrizio Candeloro [1], Francesco Gentile [4,*] and Natalia Malara [1,*]

1. Department of Experimental and Clinical Medicine, University of Magna Graecia, 88100 Catanzaro, Italy; coluccio@unicz.it (M.L.C.); gerardo.perozziello@unicz.it (G.P.); patrizio.candeloro@unicz.it (P.C.)
2. Department of Health Sciences, University of Magna Graecia, 88100 Catanzaro, Italy; pullano@unicz.it (S.A.P.); marco@vismara.info (M.F.M.V.)
3. Institute of Materials for Electronics and Magnetism (IMEM) CNR Parco area delle Scienze 37/A, 43124 Parma, Italy; nicola.coppede@gmail.com
4. Department of Electrical Engineering and Information Technology, University Federico II of Naples, 80125 Naples, Italy

* Correspondence: francesco.gentile2@unina.it (F.G.); nataliamalara@unicz.it (N.M.); Tel.: +39-096-1369-4341 (F.G. & N.M.)

Received: 30 November 2019; Accepted: 20 January 2020; Published: 22 January 2020

Abstract: A long-standing goal of nanoelectronics is the development of integrated systems to be used in medicine as sensor, therapeutic, or theranostic devices. In this review, we examine the phenomena of transport and the interaction between electro-active charges and the material at the nanoscale. We then demonstrate how these mechanisms can be exploited to design and fabricate devices for applications in biomedicine and bioengineering. Specifically, we present and discuss electrochemical devices based on the interaction between ions and conductive polymers, such as organic electrochemical transistors (OFETs), electrolyte gated field-effect transistors (FETs), fin field-effect transistor (FinFETs), tunnelling field-effect transistors (TFETs), electrochemical lab-on-chips (LOCs). For these systems, we comment on their use in medicine.

Keywords: biodevices; integration; miniaturized devices

---

1. Introduction

Theranostics is universally understood to be the use of a combination of nanoscale agents and techniques that have both diagnostic and therapeutic effects on a disease. Theranostics provides a transition from conventional medicine to a personalized and precision medicine approach.

It includes a large variety of themes including, for example, molecular imaging, personalized medicine, or pharmacogenomics that expand the field of knowledge on targeted therapies and enhance our understanding of the molecular mechanisms of drugs. In the last years, recent advances in microfluidics [1–4] and nanotechnology [5–8] gave significant support to theranostics for developing new procedures and treatments of diseases.

This review focuses on the development of strategies and potential applications of emerging theranostic nanosystems based on the transport of electroactive species (i.e., ions or electrons) at the nanoscale. Faster diagnosis and screening of diseases have become increasingly important in predictive personalized medicine as they improve patient treatment strategies and reduce cost as well as the burden of healthcare. However, the correct extraction, acquisition, and sampling of physiological signals is still unsatisfied by many of the currently available approaches; for these, the analysis of body fluids such as tears, sweat, saliva, or interstitial fluid, is heavily conditioned by the type of biomarker integrated with the electrochemical/bioelectronics sensor that performs the analysis.

The choice of the correct biomarker is a critical step that can compromise the entire process of measurement and overrule even the more advanced technology of a sensing device. The blind biomarker should exhibit the properties of high specificity and sensitivity in monitoring a disease. This adherence must be investigated in a preliminary stage, before the use of the biomarker in the device in its final configuration. Then, the aim of the (electrochemical) device is that of enhancing the sensing abilities of the biomarker in terms of limit of detection (i.e., the smallest quantity or concentration of the analyte to be detected), resolution (i.e., the smallest incremental unit of the analyte that the biosensor can detect), and sensitivity (i.e., how much the response of the system changes as the input changes). A nanoscale architecture of the device can improve the limit of detection, resolution, and sensitivity of the biosensor. This review is an account of how nanotechnology can enter the field of electrochemical transistors to impact clinical medicine. The circuit functionalities and their applications will also be addressed, with attention to current trends in the field.

For their characteristics of high sensitivity and fast response times, electrochemical biosensors may provide early diagnosis of diseases and increase the possibility of a patient's recovery. Perhaps more importantly, electrochemical biosensors are able to translate a chemical signal into an electrical signal and this enables us to detect and quantify several different molecular or cellular species in the body. Moreover, these systems can be integrated with lab-on-chips to obtain point of care (POC) analytical platforms.

In the fields of theranostics and prognosis, it is necessary to develop devices with a high impact on the detection sensitivity and specificity of the biomarker that in turn must be specific and adherent to the disease under examination. The integration between electrochemical biosensors and lab-on-chips (LOCs) to obtain POC analytical platforms is discussed with plenty of examples. We report, describe, and comment on latest generation transistors, electrochemical biosensors (fin field-effect transistors (FinFETs), tunnelling field-effect transistors (TFETs), and organic electrochemical transistors (OECTs)), and the combination of electrochemical biosensors with lab-on-chips for medical applications.

2. Fin Field-Effect Transistor (FinFET), Tunnel FET

The field-effect transistor is a type of transistor that uses an electric field to control the flow of current. A typical FET device has three terminals: source, gate, and drain. The application of a voltage to the FET's gate modifies the conductivity between the drain and the source, allowing the control of the flow of current.

One of the first examples of field-effect devices for the evaluation of ionic species was introduced by Bergveld in the 1970s and named ion sensitive field-effect transistor (ISFET) [9].

In this class of devices, the gate consists of an SiO_2 layer placed in solution, and consequently, the drain current is influenced by analyte activity by varying the potential at the gate/electrolyte interface [9,10]. Subsequently, different technologies of field-effect devices were developed to overcome the main limits imposed by FET–based devices, such as the threshold voltage drift. In chronological order the extended gate field effect transistor (EGFET) was developed by Van der Spiegel in the early 1980s [11]. The continuous scaling of planar metal-oxide-semiconductor field-effect transistor (MOSFET) evidenced come technological difficulties whereby the device can no longer be classified as a long channel MOSFET. The main short-channel effect is due to the two-dimensional distribution of potential and high electric fields in the channel region, which mainly lead to variable threshold voltage, saturation region that does not depend on drain potential, and drain current that does not depend on the inverse of channel length [12]. Subsequently a variety of different planar and non-planar topologies were investigated, such as the FinFET and the TFET [13,14]. The FinFET resulted in an attractive option for the fabrication of a non-planar device with self-aligned double-gate using a standard complementary metal-oxide semiconductor (CMOS) process [15]. Conversely, the FET is considered a promising design which guarantees immunity to subthreshold swing degradation at short channel length [16]. More recently, the development of more sophisticated organic materials led to the development of devices that are trying to replace the role of silicon, such as the organic thin-film transistors (OTFTs). Even though the OTFTs' performances are actually not comparable with inorganic ones in terms of carrier mobility, operating frequency, and subthreshold swing, the low cost and easier fabrication, as well as the possibility to chemically modify the material properties, represent a key role in the development of organic biosensors [17].

The electronic interface of a field-effect biosensors' electronic interface is a standard and/or custom designed MOSFET, which ensures long-term stability and insulation from the chemical environment to the device (where the sensing layer is generally placed) [16]. In linear region, the sensor output (i.e., drain current) is related to the analyte as follows:

$$I_{DS} = \mu C_{ox} \frac{W}{L}\left[\left(V_{Ref} - V_{th}^{*}\right)V_{DS} - \frac{1}{2}V_{DS}^{2}\right] \quad (1)$$

here W is the width and L is the length of the channel, μ is the carrier mobility, C_{Ox} is the gate oxide capacitance per unit area, and V_{Ref} and V_{DS} are the applied reference electrode and the drain-to-source voltages. V_{th}^{*} is the threshold voltage, which is related to the device and the chemical environment as follows:

$$V_{th}^{*} = V_{th} + E_{ref} + \chi_{sol} - \frac{W_M}{q} - \phi \quad (2)$$

In Equation (2), V_{th} is the threshold voltage of the field-effect device. E_{Ref} is the reference electrode potential, χ_{sol} is the dipole potential of the electrolyte, W_M is the work function of the reference electrode, q is the charge, and φ is the potential at the sensing interface [10]. In Figure 1 is reported a comparison between an EGFET and a Fin-FET device for biosensing.

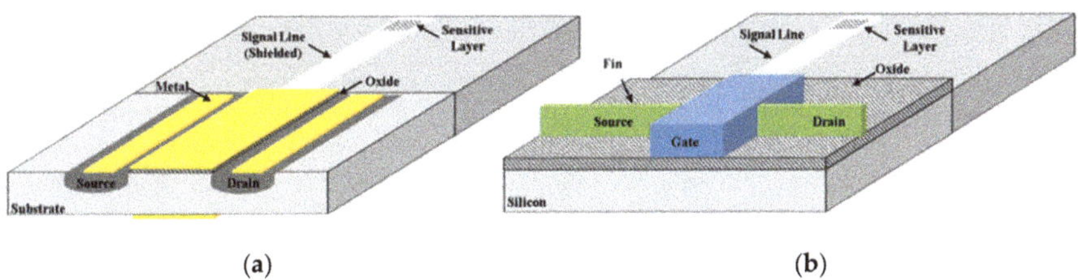

(a) (b)

Figure 1. Representative view of a device based on extended gate field effect transistor (EGFET) (**a**) and (**b**) fin field-effect transistor (Fin-FET) technology (not in scale).

In the EGFET aspect ratio W/L influences the characteristics of the devices in terms of transconductance g_m. A higher transconductance is desirable because it results in lower flicker ($1/f$) noise device. There are several advantages in FinFET such as a lower off current, a lower V_{th} due to a reduced bulk (depletion) capacitance and a very low output conductance (higher voltage gain). Conversely, FinFETs suffer from a high series resistance and thus a lower peak transconductance [18]. In both cases the literature reports that most of the developed biosensors are developed using commercial devices, evidencing how the development of biosensors over the years has been mostly oriented toward the sensing part, using off-the-shelf components because of the easier fabrication process and lower cost [19,20]. Even though they are a more recent technology, FinFETs have more recently been commercialized and thus are mature for biosensor applications [21]. One of the key metrics in the development of biosensors is their sensitivity, which through the years has been the object of intensive investigation, especially for the detection of analytes at even lower concentration. Being inherently characterized by theoretical limits other approaches were investigated instead of classical planar and non-planar geometries. The TFET is one of the most recent devices, with base conduction mechanism on the band-to-band tunneling. In this class of device, the analyte influences the tunneling barrier, and hence the tunneling current. Literature has evidenced that the use of TFET technology results in devices with improved sensitivity and reduced response time, while retaining all other advantages of FET biosensors [22–24]. FinFET technology was originally proposed as an improved technology characterized by higher sensitivity, stability, and reliability [25]. Literature reports some attempts to fabricate FinFET-based sensors for biomolecule detection such as cellular ion activities [26], pH [25,27], and the detection of avian influenza (AI) antibody [28]. Change in current was recently linked to change in gate capacitance, allowing the detection of proteins linked to early detection of diseases (e.g., streptavidin, biotin) [29]. Moreover, being a relatively recent technology, modelling tools are still under development to optimize the design phase [30].

The continuous efforts to improve the sensing performances attracted significant attention through the recent technological advancement in synthesis and deposition on high performance materials, such as graphene (i.e., nanopores, nanoribbon, reduced graphene oxide and graphene oxide), carbon nanotube, nanowires, and nanoporous materials [31–38]. Graphene is a high-performance material, recently investigated in different fields due to the availability of synthesis and mature deposition technologies, characterized by high carrier mobility and low inherent $1/f$ noise [39]. As result, different attempts at using a graphene-FET (GFET) as biosensor were reported in literature. Most of the applications are focused on low-concentration nucleic acid detection, exploiting the site-specific immobilization of probes [32]. The reported resolution of GFETs, often conjugated with metal nanoparticles (e.g., Au) can be lowered down to the pM range [40,41]. Despite the interesting sensing applications, continuous investigations are still required to improve the reduced DNA translocation dynamics and the low-frequency noise levels [32,42,43]. Other applications are concerned with protein detection, living cell and bacteria monitoring [33,44]. A commercial graphene-based biosensor is the agile biosensor chip—NTA, used overall for research purposes, allowing the immobilization of recombinant proteins.

The efforts to develop the FinFET device often lead to biosensors with performances comparable with that of other multigate or planar MOSFETs [25]. Subsequently, in order to reduce the development time, commercial FET devices are sometimes preferred.

3. Organic Electrochemical Transistor Devices

The role of organic electrochemical transistors in biomedicine is becoming increasingly relevant. OECTs are devices based on a semiconductor, conventionally named channel, that is physically placed in contact with a solution. Upon the action of an externally controlled voltage, ions of the solution are displaced and can be either injected or removed from the channel, doping or dedoping it, changing the bulk conductivity of the entire device. Alterations in the electrical characteristics of the device can be in turn correlated to the physical and chemical characteristics of the electrolyte. State of the art OECTs are mostly based on the conducting polymer poly(3,4-ethylenedioxythiophene) doped with polystyrene sulfonate (PEDOT:PSS) [45], thus OECTs share many of the characteristics of polymers such as low weight, low density, low cost, high resilience, elevated specific strength, biocompatibility, and facile deposition. Moreover, because they have a transistor architecture, OECTs feature high sensitivity, high signal-to-noise ratio, high gain [46–48], enabling amplification of weak electric signals such as those generated by biological systems.

OECTs, in fact, can operate at low voltages, in aqueous environments such as efficient ion-to-electron converters, providing an interface between the worlds of biology and electronics.

As key constituents of biosensors and analytical devices, OECTs have been used for electrophysiological recording, for bio-sensing applications and applications at the bio-interface [45,49, 50], bio-computing [51], neuromorphic engineering [52–54], as constituents in electronic bio-devices [55], and as a sensor for cells [56].

As flexible, high-sensitivity low-cost devices, OECTs have found different applications in biomedicine and biology. They have been applied as sensors for simple analytes such as hydrogen peroxide [57] and ions [47,58]. In reference [59] Seong-Min Kim and colleagues examined the long term stability of PEDOT:PSS, examined the correlations among the microstructure, composition, and device performance of PEDOT:PSS films for possible applications in the development of long-term stable implantable bioelectronics for neural recording/stimulation. In reference [60], devices based on the conducting polymer PEDOT:PSS have been demonstrated for the real-time processing and manipulation of signals from living organisms; devices with the characteristics of miniaturization and bio-compatibility with human skin have been used to analyze neurophysiological activity. In references [61–63] it is discussed how similar OECTs can be used as sensing interfaces of cells and systems of cells. OECTs can very practically determine the physiological conditions of living cells, follow the processes at the basis of their life cycle, including reproduction processes, and track their transition to apoptosis [63].

Moreover, OECTs can be used as electronic switches or components of logic gates, also in interaction with biological interfaces, creating a multiple interactive logic that is perfectly suited to talk with living systems [64]. The in vivo monitoring of biological-driven phenomena is likewise heavily investigated, including, for example, enzymatic interactions [65] useful to detect metabolites relevant in in the biological processes and function of cells and organs, such as lactate or glucose, which are indicative of the physiological conditions of a patient [66]. Finally, a wide range of applications is still open in the detection of biomolecules based on specific antigens, which are crucial in the in vivo early diagnosis of bacteria and specific illnesses [67].

The typical configuration of an OECT is reported in Figure 2. An electrolyte, i.e., a solution containing electroactive species, is contacted to the device with three electrodes: the gate, the drain, and the source. The gate is the reference electrode that is directly connected to the electrolyte. Instead, the drain and the source are bridged together by the conductive polymer channel (Figure 2a). Upon application of a voltage between the gate and the source (V_{gs}) and the drain and the source (V_{ds}), ions in the electrolyte are propelled towards the polymer channel, penetrate into the channel, and generate a current I_{ds} that flows from the drain to the source (Figure 2b). Thus, the current I_{ds} is the typical output of an OECT device. The reaction between the ions and the polymer in the channel is described by the following equation:

$$\text{PEDOT}^+ : \text{PSS}^- + \text{M}^+ + e^- \rightarrow \text{PEDOT} + \text{M}^+ : \text{PSS}^- \tag{3}$$

where M^+ represents the cations. The presence of cations in the PEDOT : PSS film depletes the number of available carriers, reducing the source–drain current I_{ds}. Thus, I_{ds} is modulated by the inflow or outflow of ions—from the electrolyte to the channel—and values of current measured by the device can be indicative of the concentration, size, and charge of the specie initially dispersed in solution, and of the geometrical characteristics of the system. The form of the I_{ds} current is similar to response of a first order system: values of current are a function of time and increase from a reference value (minimum background current, no flux) to a steady state value (maximum value of current, constant regime) (Figure 2c). The signal can be modelled by an exponential function of the type $I_{ds} \sim m\left(1 - e^{-t/\tau}\right)$, where t is time, and m and τ are the modulation and time constants of the system. The modulation is the signal increment normalized to its initial value, $m = \left(I_{ds}^{fin} - I_{ds}^0\right)/I_{ds}^0$; it is proportional to the strength of the signal. The time constant is the time after which the signal attains 67% of its steady state value, $I_{ds}(\tau) \sim 0.67\, I_{ds}^{fin}$; it is indicative of the inertia of the system. Remarkably, using mathematical models described elsewhere [68] m and τ can be associated with the diffusivity, charge, concentration, and other physical, chemical and geometrical characteristics of a system, so that the information encoded in the I_{ds} can be broken to extract the characteristics of the system in analysis. Typical I_{ds} values and increments fall in the 0–5 mA range. The values of V_{ds}, instead, are controlled by the operator and are typically varied in discrete increments in the 0–1 V range.

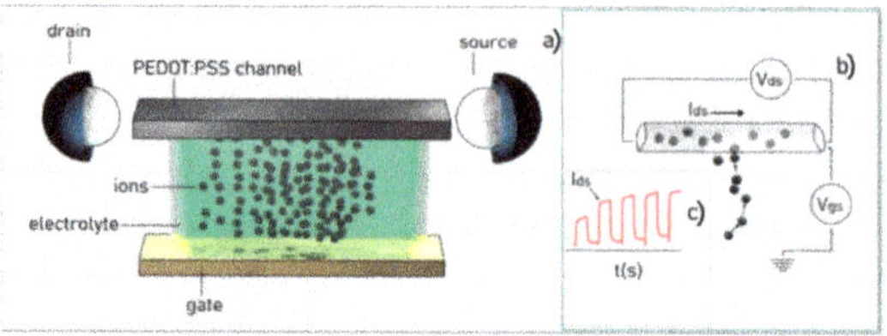

Figure 2. Scheme of a conventional organic electrochemical transistor (OECT) device, in which a solution is connected to the device through the gate, source and gate electrodes, and a conductive polymer channel (**a**). Upon application of an external voltage at the gate and the drain (**b**), a current of ions flows to the source generating a continuous function of time (**c**).

In this scheme, the electrolyte is contained in a channel, a chamber, or a reservoir; the interface of the solution with the external regions of the device is a flat surface with zero curvature. This automatically implies that the motion of ions in the system to the active sites of the device is driven by diffusion: the process is inefficiently controllable by the outside. By nanostructuring of the surface of the polymer channel, one can make the surface super-hydrophobic. Super-hydrophobicity prevents wetting of the surface, allowing a drop of solute positioned on that surface to maintain its originating spherical shape. The curvature of the drop can be modulated by tailoring the geometry of the super-hydrophobic surface and by regulating the amount of the solute partitioned in each drop. Thanks to a non-zero curvature, convective Marangoni flows arise in the drop (Figure 3a).

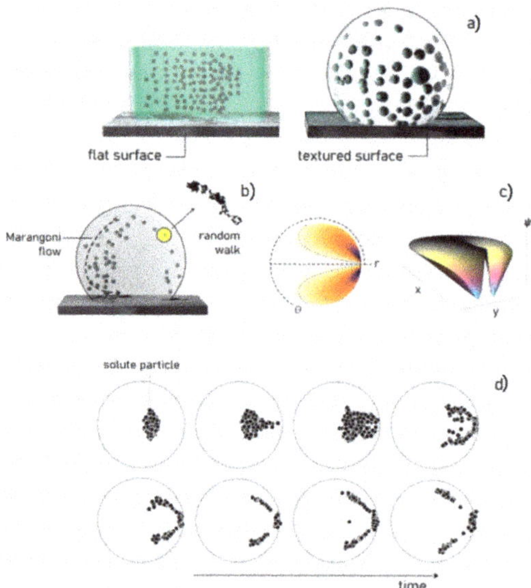

Figure 3. Nanoscale modification of a surface can make that surface super-hydrophobic (**a**). In a drop on a super-hydrophobic surface, the motion of particles is determined by the combination of diffusion and Marangoni convective flows (**b**). The inset reports a graphical representation of the potential functions ψ that describe the velocity field within a spherical drop (**c**). The displacement of particles in a drop can be determined from the velocity field using the Langevin equation and a numerical scheme (**d**).

Depending on the value of curvature, the size of the drop, and the gradient of temperature between the substrate and the drop, the intensity of the convective fields can be equal, or greater or less, than the intensity of diffusion. Thus, convection represents the additional degree of freedom introduced in the system [69], and the competition between convection and diffusion drives the solute species on predefined spots. The velocity field developed within the drop owing to Marangoni flows is derived as the derivative of the stream functions $\psi(r,\theta)$ with respect to the coordinates r and θ. The stream functions are potential functions that describe the characteristics of a fluid flow, they have been originally derived by Tam and collaborators and [70] read:

$$\psi(r,\theta) = -\frac{1}{8}\left(1-r^2\right)\left[1 + r\cos\theta - \frac{1-r^2}{\left(r^2+1-2r\cos\theta\right)^{\frac{1}{2}}} + \sum_{n=2}^{\infty} \frac{(n-1)-2(n-1)Bi}{(2n-1)((n-1)+Bi)} r^n \left(P_{n-2}\cos\theta - P_n\cos\theta\right)\right], \tag{4}$$

where (r,θ) is the position of a point in the drop in polar coordinates, Bi is the Biot number, and P_n is the Legendre polynomial of order n. The corresponding velocity field v is then derived as:

$$v_r = -\frac{1}{r^2 sin\theta}\frac{\partial\psi}{\partial\theta},\ v_\theta = \frac{1}{r\ sin\theta}\frac{\partial\psi}{\partial r}, \tag{5}$$

that can be used, in turn, in the Langevin equation [71–73] to find the distribution of a trace in the drop:

$$m\frac{\partial \boldsymbol{u}}{\partial t} = 6\pi\mu a\left(K_p\boldsymbol{u} - K_f\boldsymbol{v}\right) + F_e + F_b. \tag{6}$$

In Equation (6) u is the unknown velocity vector for the particle, v is the unperturbed fluid velocity, m and a are the mass and radius of the solute particulates. Moreover, F_e is the electrostatic force, F_b is the Brownian force that depends on the temperature as $F_b \propto \sqrt{T}$, K_p and K_f account for the hydrodynamic hindrance of the system, and t is time. Equation (6), solved using a numerical scheme, allows us to determine how a solute propagates in a super-hydrophobic drop because of convection and diffusion.

Inspired by Lotus leaves and by nature, super-hydrophobic surfaces have been reproduced using combinations of nano-fabrication techniques [74,75]. Typically, the artificial analogue of a super-hydrophobic surface is an array of microsized pillars, where the size (d), spacing (δ), and height (h) of the pillars in the array can vary over large intervals. The upper surface of the pillars contains, in turn, details at the nanoscale, and the combined effects across length-scales cause the surface to be super-hydrophobic [76–78]. The pillars of those surfaces are often made out of silicon, nano-machined using reactive ion etching techniques, or of polymers, created using optical or electron beam lithography techniques [79]. Fluorinated polymers with low friction coefficients, nano-porous silicon, or nano-rough materials with low surface energy densities, can be deposited on the pillars representing the second-level roughness of the hierarchical nanomaterial device [80]. The contact angle of a drop on similar super-hydrophobic surfaces can be predicted using the celebrated model of Cassie and Baxter [74]:

$$cos(\vartheta^c) = -1 + \phi\ cos(\vartheta), \tag{7}$$

where ϑ is the contact angle of the drop on the surface without texture, ϑ^c is the contact angle on the surface with the texture, and ϕ is the solid fraction of the surface. Typical design values of d and δ are $d = 10\ \mu m$ and $\delta = 20\ \mu m$, so that $\phi = \pi/4(d/\delta)^2 \sim 0.087$. With these values of d, δ and ϕ, any originating contact angle $\vartheta > 60°$ will lead to final contact angles $\vartheta^c > 150°$, i.e., a super-hydrophobic surface. For this combination of d and δ, the height of the pillars should be chosen such that $h \geq 20\ \mu m$, in that ratio of h to δ is greater than one, $h/\delta > 1$, to assure stable adhesion of the drop on the surface [81].

Motivated by the need of new sensor devices with higher sensitivity, higher accuracy, and increased selectivity with respect to available approaches, beginning in 2014 some of the authors of this paper started to nanostructure OECTs with the aim to harness their functionalities [82]. Starting from conventional OECT devices, we modified the geometry of those devices at the nanoscale, and created a new class of bio-devices that we called surface enhanced organic electrochemical transistors

(SeOECTs) [69]. SeOECTs are a third generation of organic thin film transistors, in which the electrolyte medium is an active part of the device gating and the surface micro and nanostructure enhances the properties of the electrochemically active conductive polymer. In SeOECTs, a 3-dimensional design and topographical modification of the surface enables selectivity, enhances sensitivity, and enables the detection of multiple analytes in very low abundance ranges.

SeOECTs are based on the fine tailoring of surface microstructure and nano structure. The device comprises arrays of super-hydrophobic micro-pillars, functionalized with a conductive PEDOT:PSS polymer sensitive to the ionic strength of the electrolyte. Each pillar has a diameter of 10 μm and a height of 20 μm. The pillars are positioned on the substrate to form a non-periodic lattice (Figure 4a). A similar non-uniform tiling of pillars generates a system of radial forces that recalls the drop to the center of the lattice for automatic sample positioning. Some of the pillars are individually contacted to an external electrical probe station for site selective measurement on the sample surface (Figure 4b,c); they incorporate nano-gold contacts with sub-micron reciprocal distance that generate enhanced and localized electric fields (Figure 4d–h). Due to the microstructure of the device, the device is super-hydrophobic with contact angles up to 165° (Figure 4i). The device takes advantage of a combination of scales to resolve, identify, and measure complex biological mixtures. At the micro-scale, arrays of super-hydrophobic micro pillars enable manipulation and control of biological fluids. At the nano-scale, some pillars are modified to incorporate nano-electrodes for time and space resolved analysis of solutions.

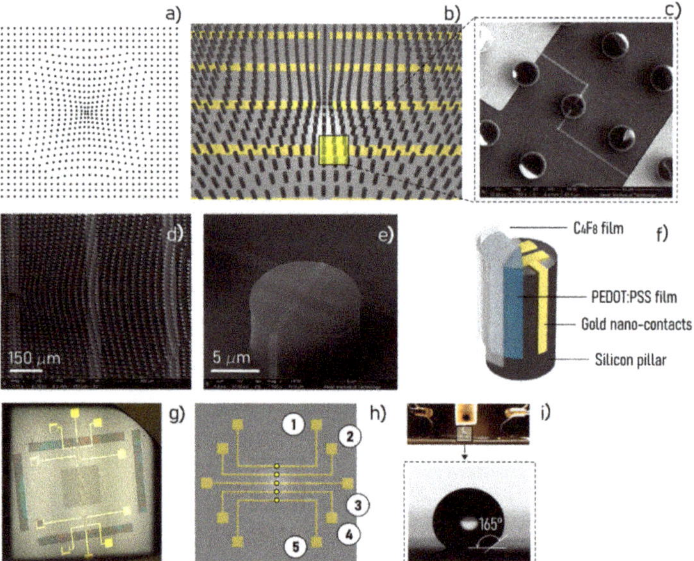

Figure 4. Surface enhanced organic electrochemical transistors (SeOECT) devices are based on a non-periodic array of micro-pillars (**a**). Some of those pillars are contacted through circuits to an external electric probing station (**b**) and are equipped with nanoelectrodes for site selective measurement of the ionic current: each of those pillars is named sensor (**c**–**f**). The device imaged with a camera lens, the distance between the parallel gold circuits is 1 cm (**g**). The sensors are placed in line on the device, symmetrically with respect to the center of the device (**h**). Due to the characteristics of the substrate, during operation the liquid sample maintains a spherical shape (**i**).

SeOECTs are obtained by the superposition of different layers as explained in detail in reference [69].

To perform a measurement, a liquid sample is positioned on the device. Due to the super-hydrophobic characteristics of the surface, the sample takes a quasi-spherical shape (drop) and is automatically centered on the device. Then, the device is driven by an externally applied voltage, ranging between 0 and 1 volt. Upon the application of the voltage, a current of ions I_{ds} flows from the sample through the circuit and is measured by the points of measurements (sensors) on the device. Ions in the sample drop are transported by buoyancy and Marangoni flows that originate in the drop because of its curvature. Since the motion of ions—under the combined effect of convective flows and electric field—is directly proportional to the charge, directly proportional to the diffusion coefficient, and inversely proportional to the size, the process achieves the migration and spatial separation of species in solution. Arrays of sensors, spatially positioned on the device, can resolve this separation in space and time. Thus, the device measures ionic current transients at different positions on the substrate and for different values of voltage. While the information content of the solution is mapped into a whole set of variables, statistical techniques of analyses can be used to decode such information and determine the characteristics of target molecules.

The state of the system in a specific configuration is a point in the $m - \tau$ plane. Samples with different characteristics are placed in different regions of the diagram (Figure 5a). Thanks to this graphical representation, it is possible to operate sample separation, clustering, and classification (Figure 5b). The separation can be optimized if one considers sensors positioned at opposite extremes of the device or high values of voltage, for which the convective transport effects are amplified and the differences between species with different charge and size are maximized (Figure 5c). Points in the $m - \tau$ diagram can be parametrized by voltage (Figure 5d) or by time (Figure 5e). In both cases, data processing and analysis generate trajectories, the shapes of which are indicative of the time evolution and of the characteristics of the system (Figure 5f).

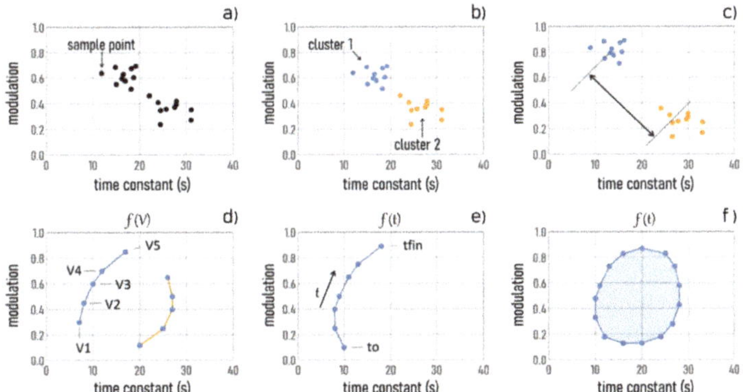

Figure 5. The response of the SeOECT devices is described by the sole variables modulation and time constant (**a**). The scatter plot of modulation against time constant may be indicative of differences between samples and can be used to operate sample separation, clustering, and classification (**b**). Separation between species can be improved by setting high voltage values and using sensors positioned at the border of the drop, where Marangoni flows are maximized (**c**). The modulation and time constant variables can be parametrized by voltage (**d**) and by time (**e**). The form of these trajectories can be indicative of the time evolution of the system (**f**).

SeOECT devices have been used to evaluate tumors [83]. Since cancerous states are associated with an altered protonation state of the intra/extracellular microenvironment, one can estimate the onset and progression of a cancerous disease from a measurement of the ionic content of a blood-derived cell culture. We used SeOECTS to evaluate potential perturbation of protein protonation state (i.e., charge) of cell secretome in the extracellular compartment in vitro. We applied the analysis to the conditioned medium of blood culture after a short-time expansion, derived from patients with, without, and suspected of cancer. Using data from the bio-chip and statistical techniques of analysis, we developed algorithms that segregated tumor patients from non-tumor patients. For the ~30 patients across two independent cohorts, the method identified tumor patients with high sensitivity and 93% specificity [83].

4. Combined Electrochemical Biosensor and Lab-on-Chip

The sensing platform is then a crucial point for a specific application of LOCs, consisting of a sort of recognition element required for the capture of the target. The high affinity between antigen–antibody (Ag–Ab) or protein–aptamer makes them largely applied for biosensing design. Nanotechnologies offer support as electrochemical devices because the nanomaterials employed, from silicon to graphene or graphene oxide to carbon nanotubes or to metal nanoparticles, are characterized by excellent electrical, mechanical, and, generically, physical/chemical properties, which, combined with the novel nanotechnology techniques (lithography, metal deposition, plasma treating), allow the realization of appropriate architectures with the appropriate functionalities too. In particular the optical and electrical properties of the sensing nanodevices are related to their materials as well as to their geometry, and an optimal combination of them can amplify the signals coming from the analytes. Consequently, high-sensitivity analyses of biomarkers (usually cellular biomarkers or biomolecules) become possible adopting nanodevices with different techniques (e.g., cyclic voltammetry (CV), electrochemical impedance spectroscopy (EIS), IR, or Raman spectroscopy) [84,85]. Reference [7] reports one of the first examples of a sensor able to capture circulating tumor cells (CTCs), which could be used to study tumor staging, to guide the study of the risk of recurrence. Folic acid (FA) was selected as transducer molecule, because CTCs characterized by high expression of folate receptor (FR) and, consequently, evidencing also 5-methylcytosine-positive nuclei, are potentially dangerous for their dissemination power in healthy tissue. The FA surface can trap cancer-cell-expressing FR, encouraging the attack and the growth of CTC subsets with a higher content of methylated genomic DNA, with notable consequences on tumor prevention strategies and on prognosis definition.

Microfluidic paper-based analytical devices (μPADs) represent a technology of hydrophilic/hydrophobic micro-channel networks and associated analytical devices for development of portable and low-cost diagnostic tools that improve point of care testing (POCT) and disease screening [86] involving the specific detection of biomolecules. They have the ability to perform laboratory operations on micro-scale, using miniaturized equipment, and can be fabricated by using 2-D [86–88] or 3-D [89,90] methods to transport fluids in both horizontal and vertical dimensions, depending on complexity of the diagnostic application. The principal techniques in the literature for fabrication of paper-based microfluidic devices include: wax printing [90], inkjet printing [91], photolithography [92], flexographic printing [93], plasma treatment [94], laser treatment [95], wet etching [96], screen printing [97], and wax screen printing [98]. The μPADs can be used with the naked eye for qualitative testing but can also be used as quantitative assays based on specific detection methods. The choice of the method to detect the binding events that occur on a transducer surface depends on the type of biomarker. The detection methods currently used for sensitive measurements with low detection limits are represented by colorimetry, electrochemistry, fluorescence, chemiluminescence (CL), electrochemiluninescence (ECL), and photoelectrochemistry (PEC). Moreover, the surfaces can be modified to impart high selectivity to the binding of the target analyte, which is desirable in complex biological samples. In fact, sensitivity and specificity of the μPADs can be enhanced through a combination use of the reaction mechanisms categorized into biochemical, immunological, and molecular detections, transforming the device in a multiplexed testing [99–101]. However, chemical amplification or multiplex procedures are often disadvantaged in the μPADs as they require expensive reagents, multiple steps that must be performed by the end user, and complex protocols for the interpretation of the final data.

5. Future Trends of Combined Electrochemical Biosensor and Lab-on-Chip

The μPADs promise to meet the critical needs of rapid analytical tests in the diagnostic area. These devices represent the diagnostic field, a platform for a wide variety of chemical and biochemical reactions and detection patterns that can be used to assess the health status of the general population. They are useful for people who must reach very distant health facilities (Figure 6) [102]. These devices can also be used to monitor intoxications in occupational medicine. Finally, their ever-increasing application in the qualitative assessment of foods is beginning. Further research is needed to address several common challenges, such as the poor reproducibility, the need of high detection limits, the inadequate specificity, and the risk of a subjective interpretation of data. Most μPADs successfully address most of these challenges through the association of a machine learning procedure based on "kernel machines." Kernel machines have considerable appeal in the machine learning research community due to a combination of conceptual elegance, mathematical tractability, and state-of-the-art performance [102], and, applied on Android smartphones for image processing and paper-based devices, they are able to solve many of the several common challenges mentioned.

Figure 6. Flowchart of the use of an Android smartphone for image processing and paper-based devices. In this example, the test is performed at home, the data collected from the paper-based device through the Android application, and adequate software is sent to qualified medical personnel to support the management of the results obtained.

The calculations of the limit of detection (LOD) and limit of concentration (LOC) for the combination of Android smartphones with image processing and paper-based devices are considered the frontier for their use. Despite its limitations, the interesting combination of the μPAD and Android applications provides a coherent and objective analysis of colorimetric data without the need of complicated interpretation data methods, and consequently it represents a significant milestone in the current and future development of ePADs for clinical diagnostics

Finally, attracting increased attention from the research community is the development of mobile device-based healthcare as a revolutionary approach for monitoring medical conditions. More systems have computerized traditional clinical tests to design functions and interactions of smartphone-based rehabilitation systems [103] regarding diseases like as stroke and cardiac failure [104]. Moreover, sophisticated platforms were developed for a better-targeted cancer therapy and improved follow-up care, to make the care process more effective in terms of clinical outcome. On the other hand, there is also the need to develop the μPAD for personalized toxicity studies. Therefore, the future trends on theranostic applications of the μPAD will be developed from the bench-to-bedside and updated to produce patient-friendly analytical assays [105].

6. Medical Clinical Applications

In cancer, the next generation of POC will probably be represented by 2-D material-based electrochemical biosensors/sensors [106] such as electrochemical apparatus [107], lateral flow assays (LFAs) [108], or paper-based colorimetric solutions [109]. The main biomarkers relevant in this field are nucleic acids such as mRNA and DNA; proteins such as antigens, enzymes, and peptides; some small molecules such as the reactive species of oxygen and nitrogen; and, notably, the protonation state [83].

Due to the chaotic nature of this medical condition, multiple marker solutions can fit the clinical needs better. Aptamers [110] are artificial oligonucleotides selected through a Systematic evolution of ligands by exponential enrichment (SELEX) procedure. They are gaining appreciation as ultra-specific, stable probes. Their use spans therapeutics to diagnostics, where they can be used as biomarker surrogates or in so-called aptahistochemistry, an evolution of immunohistochemistry [111].

In this section of the review some examples related to the more diffuse cancer types and infectious diseases are reported. Cancer-related applications are summarized in Table 1, infectious disease applications are summarized in Table 2.

In breast cancer aptamers are gaining momentum, and they find usage in therapy, as well as in the detection of diagnostic and prognostic markers such as aberrant HER-2 forms [112,113], α-estrogen receptor status [114], vascular endothelial growth factor (VEGF) [115], osteopontin [116], Michigan Cancer Foundation-7 (MCF-7) cells [117], anterior gradient homolog 2 (AGR-2) protein [118].

In lung cancer, one of the main issues is early diagnostics and effective screening. The problem is to find a proper balance between sensitivity and specificity, and complexity, time, and expenses. According to a recent review by Roointan and colleagues [119], the main biosensor approaches in lung cancer early diagnosis are electrochemical, optical, and piezoelectric (mass-based). The best clinical, analytical, and technological performances are achieved by electrochemical sensors. The main biomarkers sensed by those instruments, are: VEGF165 [120,121], EGFR [122], Annexin II and MUC5AC [123], HIF-1α [124], NADH levels [125].

In colorectal cancer (CRC) the main early diagnostic biomarker is the fecal occult blood test (FOBT) [126,127]. The most advanced FOBT tests commonly available on the market are immunochemical [128,129]. These can be further divided into qualitative and quantitative [130]. One of the main advantages of the immunological-based approach is that patients are allowed to stay on a regular diet without the need to stop drugs known to interfere with the guaiac-based FOBT [128].

Instead of looking for occult blood in stools, it is possible to approach early diagnosis looking for common genetic aberrations commonly found in CRC, such as K-Ras, adenoma polyposis coli (APC), p53, and microsatellite instability. Moreover, the novel DNA tests comprise epigenetic analysis of methylated genes for vimentin, secreted frizzled-related protein 2 (*SFRP2*), bone morphogenetic protein 3 (*BMP3*), N-Myc downstream-regulated gene 4 protein (*NDRG4*), and tissue factor pathway inhibitor 2 (*TFPI2*), using as analytical matrix feces or venous blood. Another promising target for early diagnosis and screening are fecal miRNAs [129–138].

A totally disruptive approach to FOBT is an ingestible micro-bio-electronic device (IMBED) based on environmentally resilient biosensor bacteria for in situ biomolecular detection, coupled with miniaturized luminescence readout electronics that wirelessly communicate with an external device. According to the authors, gut biomolecular monitoring could be more precise and faster than any other laboratory methods [139].

Another biomarker relevant in GI tract neoplasms is sarcosine. It is not only associated with CRC and stomach cancer, but also with prostate cancer and neurodegenerative disorders. Analytically it is relevant to human pathology in the food and fermentation industry. Many biosensor-based approaches have been tried to measure this analite: amperometric biosensors, potentiometric sarcosine biosensors, impedimetric sarcosine biosensors, photoelectrochemical (PEC) biosensors, and immunobiosensors. All these methods have been recently reviewed by Pundir at al. [140].

In the infectious disease market, many different tests received the Clinical Laboratory Improvement Amendments (CLIA) waivers that enable POC use [141]. Most POC rapid tests use lateral flow immunoassay (LFIA) technology, a limited number of POC diagnostics utilize molecular approaches. One of the most interesting molecular methods is nicking enzyme amplification reaction (NEAR) [142], an isothermal nucleic acid amplification. Back in 2015, the FDA approved the Alere i influenza A & B test [143] based on NEAR technology, which is an isothermal DNA amplification technique. Later also a test for group A Streptococcus (GAS) that uses throat swabs as samples was CLIA waived [144], and one for respiratory syncytial virus (RSV) [145].

Recently some reviews summarized monographically the state of the art in POC testing of common conditions. Kozel and collaborators [141] provided data for cryptococcal antigen meningitis and malaria. Grebely and colleagues in 2017 reviewed the offer for HCV (Hepatitis C Virus) POC testing [146]; Gaydos et al. *Trichomonas vaginalis* [147]; Hurt et al. HIV (Human Immunodeficiency Virus) [148]; Kelly and his group *Chlamydia trachomatis* [149]; Basile and collaborators in 2018 reviewed POC testing for respiratory viruses [150]; and Nzulu and colleagues in 2019 the one for *Leishmania* [151].

Concerning other experimental techniques in infection diagnostics and specific biosensor approaches, refer to the work by Datta et al. [152].

There are two other major chapters in infectious disease relevant for this review, apart from pure diagnosis: antibiotic susceptibility testing [153,154] and sepsis early diagnosis [155–158].

Table 1. Clinical application in oncology.

Device	Disease	Reference
RNA aptamers anti HER2	Breast cancer	[110]
Nanotube-wrapped anti-HER2 protein aptamers	Breast cancer	[113]
DNA aptamer anti ERα	Breast cancer	[114]
Upconversion nanoparticles DNA aptasensor for VEGF	Breast cancer	[115]
Electrochemical aptasensor for osteopontin	Breast cancer	[116]
MCF-7 aptamer-functionalized magnetic beads and quantum dots based nano-bio-probes	Breast cancer	[117]
G-quadruplex structured DNA aptamer against AGR-2	Breast cancer	[118]
Electrochemical detection of VEGF based on Au–Pd alloy-assisted aptasensor	Lung cancer	[120]
Electrochemical aptasensor for VEGF	Lung cancer	[121]
EGFR DNA sandwich-type electrochemical biosensor	Lung cancer	[122]
Amperometric immunosensors for Annexin II and MUC5AC	Lung cancer	[123]
Amperometric sensing of HIF1α	Lung cancer	[124]
Amperometric sensor for NADH (nicotinamide adenine dinucleotide) using activated graphene oxide	Lung cancer	[125]
Volatile organic compounds (VOC) (Different methods)	GI (GastroIntestinal)-tract cancer	[126]
Pyruvate kinase isoenzyme type M2 (M2-PK) (Different methods)	GI-tract cancer	[127]
Ingestible micro-bio-electronic device (IMBED) and miniaturized luminescence readout electronics, wirelessly communicating with an external device	GI-tract cancer	[139]
Sarcosine (Different methods)	Colorectal, prostate, and stomach cancer; Alzheimer, dementia, sarcosinemia	[140]

Table 2. Clinical application in infectious diseases.

Device	Disease	Reference
Isothermal nucleic acid amplification test	Influenza	[142,143]
Isothermal nucleic acid amplification test	Group A beta-hemolytic streptococcus	[142,144]
Isothermal nucleic acid amplification test	Respiratory syncytial virus	[141,145]
Different methods (review)	Hepatitis C	[146]
Different methods (review)	*Trichomonas vaginalis*	[147]
Different methods (review)	HIV	[148]
Different methods (review)	Urogenital *Chlamydia trachomatis*	[149]
Different methods (review)	Viral respiratory tract infections	[150]
Loop-mediated isothermal amplification (LAMP)	*Leishmania* spp	[151]
Gene Specific DNA Sensors	Pathogenic Infections	[152,154]
Different methods (review)	Antibiotic-Susceptibility Profiling	[153,154]
Different methods (review)	Sepsis (Lactate)	[155,156]
Microfluidic biochip	Sepsis (CD64)	[157,158]

7. Conclusions

The size of the global market for sensors is expected to increase in the near future due to the (i) growing demand for devices that meet the needs of early and reliable analysis of diseases and (ii) the fast pace at which technology is developing, providing products that satisfy those needs, with the additional characteristics of low energy consumption, eye-catching design, and ease of use. Here, we have reported on the most up-to-date nanotechnology prototypes in the field of sensor devices. Despite the frontier technology many of these devices possess, their functionality is often compromised by a lack of care and strategy in the pre-analytical and device-maintenance phases. In this context it is important, for example, to identify the optimal operation-interval of the sensor and tune this interval such that it matches with the range of values of the biological sample signal that are clinically relevant. Such a range, in turn, depends on the characteristics of the biological sample, or matrix (tissue position and blood/urine collection), and on the degree of advancement of the disease. This is to say that the optimization and use of a device in biomedicine is not simple and requires cooperation across disciplines and a multidisciplinary approach to ensure the right technology is adopted in the right conditions and at the right time. The many interesting contributions to biomedical nanoelectronics that we have reviewed in this paper are examples of how technology has developed to meet the needs of medicine. The nature of the problems that a new technology has to face is bifold: on one side, it has to solve specific scientific problems, and on the other side it has to adapt such a solution at the interface with medicine. Thus, it is likely that the evolution of the field of nanoelectronics in the near future will be guided by the definition of new problems in medicine.

Author Contributions: Conceptualization, N.M. and F.G.; writing—original draft preparation, N.M., F.G., M.L.C., S.A.P., M.F.M.V., N.C.; writing—review and editing, S.A.P., N.C., M.L.C.; visualization, P.C., G.P.; supervision, F.G. and N.M. All authors have read and agreed to the published version of the manuscript.

Funding: This research received no external funding.

Conflicts of Interest: The authors declare no conflict of interest.

References

1. Perozziello, G.; La Rocca, R.; Cojoc, G.; Liberale, C.; Malara, N.; Simone, G.; Candeloro, P.; Anichini, A.; Tirinato, L.; Gentile, F.; et al. Microfluidic Devices Modulate Tumor Cell Line Susceptibility to NK Cell Recognition. *Small* **2012**, *8*, 2886–2894. [CrossRef] [PubMed]
2. Perozziello, G.; Simone, G.; Candeloro, P.; Gentile, F.; Malara, N.; La Rocca, R.; Coluccio, M.L.; Pullano, S.A.; Tirinato, L.; Geschke, O.; et al. A Fluidic Motherboard for Multiplexed Simultaneous. *Micro and Nanosystems* **2010**, *2*, 4. [CrossRef]
3. Perozziello, G.; Catalano, R.; Francardi, M.; Rondanina, E.; Pardeo, F.; De angelis, F.; Malara, N.; Candeloro, P.; Morrone, G.; Di Fabrizio, E. A microfluidic device integrating plasmonic nanodevices for Raman spectroscopy analysis on trapped single living cells. *Microel. Eng.* **2013**, *11*, 314–319. [CrossRef]
4. Simone, G.; Malara, N.; Trunzo, V.; Renne, M.; Perozziello, G.; Di Fabrizio, E.; Manz, A. Galectin-3 coats the membrane of breast cells and makes a signature of tumours. *Mol. Biosyst.* **2014**, *10*, 258–265. [CrossRef]
5. Coluccio, M.L.; Gentile, F.; Das, G.; Nicastri, A.; Perri, A.M.; Candeloro, P.; Perozziello, G.; Proietti Zaccaria, R.; Gongora, J.S.; Alrasheed, S.; et al. Detection of single amino acid mutation in human breast cancer by disordered plasmonic self-similar chain. *Sci. Adv.* **2015**, *1*. [CrossRef]
6. Perozziello, G.; Candeloro, P.; Gentile, F.; Nicastri, A.; Perri, A.M.; Coluccio, M.L.; Adamo, A.; Pardeo, F.; Catalano, R.; Parrotta, E. Microfluidics & nanotechnology: Towards fully integrated analytical devices for the detection of cancer biomarkers. *R. Soc. Chem. RSC Adv.* **2014**, *4*, 55590–55598.
7. Malara, N.; Coluccio, M.L.; Limongi, T.; Asande, M.; Trunzo, V.; Cojoc, G.; Raso, C.; Candeloro, P.; Perozziello, G.; Raimondo, R.; et al. Folic Acid Functionalized Surface Highlights 5-Methylcytosine-Genomic Content within Circulating Tumor Cells. *Small* **2014**, *10*, 4324–4331. [CrossRef]
8. Pullano, S.A.; Greco, M.; Corigliano, D.M.; Foti, D.P.; Brunetti, A.; Fiorillo, A.S. Cell-line characterization by infrared-induced pyroelectric effect. *Biosens. Bioelectron.* **2019**, *140*. [CrossRef]
9. Bergveld, P. Development of an Ion–Sensitive Solid–State Device for Neurophysiological Measurements. *IEEE Trans. Biomed. Eng.* **1970**, *17*, 70–71. [CrossRef]
10. Luz, R.A.S.; Rodrigo, M.; Crespilho, I.; Crespilho, F.N. Nanomaterials for Biosensors and Implantable Biodevices. In *Nanobioelectrochemistry: From Implantable Biosensors to Green Power Generation*; Springer: Berlin/Heidelberg, Germany, 2013; pp. 27–28. ISBN 978-3-642-29249-1.

11. Van der Spiegel, J.; Lauks, I.; Chan, P.; Babic, D. The extended gate chemically sensitive field effect transistor as multi–species microprobe. *Sens. Actuators* **1983**, *4*, 291–298. [CrossRef]
12. Rajan, N.K.; Brower, K.; Duan, X.; Reed, M.A. Limit of detection of field effect transistor biosensors: Effects of surface modification and size dependence. *Appl. Phys. Lett.* **2014**, *104*. [CrossRef]
13. William, M.R.; Gehan, A.J.A. Silicon surface tunnel transistor. *Appl. Phys. Lett.* **1995**, *67*, 494–496.
14. Hisamoto, D.; Kaga, T.; Kawamoto, Y.; Takeda, E. A fully depleted lean-channel transistor (DELTA)-a novel vertical ultra thin SOI MOSFET. In Proceedings of the International Technical Digest on Electron Devices Meeting, Washington, DC, USA, 3–6 December 1989; pp. 34.5.1–34.5.4.
15. Jurczak, M.; Collaert, N.; Veloso, A.; Hoffmann, T.; Biesemans, S. Review of FINFET technology. In Proceedings of the 2009 IEEE International SOI Conference, Foster City, CA, USA, 5–8 October 2009.
16. Choi, W.Y.; Lee, W. Hetero-Gate-Dielectric Tunneling Field-Effect Transistors. *IEEE Trans. Electron Devices* **2010**, *57*, 2317–2319. [CrossRef]
17. Kergoat, L.; Piro, B.; Berggren, M.; Horowitz, G.; Pham, M.C. Advances in organic transistor-based biosensors: From organic electrochemical transistors to electrolyte-gated organic field-effect transistors. *Anal. Bioanal. Chem.* **2012**, *402*, 1813–1826. [CrossRef]
18. Yin, L.T.; Chou, J.C.; Chung, W.Y.; Sun, T.P.; Hsiung, S.K. Study of indium tin oxide thin film for separative extended gate ISFET. *Mater. Chem. Phys.* **2001**, *70*, 12–16. [CrossRef]
19. Chi, L.-L.; Chou, J.-C.; Chung, W.-Y.; Sun, T.-P.; Hsiung, S.-K. Study on extended gate field effect transistor with tin oxide sensing membrane. *Mater. Chem. Phys.* **2000**, *63*, 19–23. [CrossRef]
20. Subramanian, V.; Parvais, B.; Borremans, J.; Mercha, A.; Linten, D.; Wambacq, P.; Loo, J.; Dehan, D.; Gustin, C.; Collaert, N.; et al. Planar Bulk MOSFETs Versus FinFETs: An Analog/RF Perspective. *IEEE Trans. Electron Devices* **2006**, *53*, 3071–3079. [CrossRef]
21. Pullano, S.A.; Critello, C.D.; Mahbub, I.; Tasneem, N.T.; Shamsir, S.; Islam, S.K.; Greco, M.; Fiorillo, A.S. EGFET-Based Sensors for Bioanalytical Applications: A Review. *Sensors* **2018**, *18*, 4042. [CrossRef]
22. Pullano, S.A.; Tasneem, N.T.; Mahbub, I.; Shamsir, S.; Greco, M.; Islam, S.K.; Fiorillo, A.S. Deep Submicron EGFET Based on Transistor Association Technique for Chemical Sensing. *Sensors* **2019**, *19*, 1063. [CrossRef]
23. Ahn, J.-H.; Choi, S.-J.; Han, J.-W.; Park, T.J.; Lee, S.Y.; Choi, Y.-K. Double-Gate Nanowire Field Effect Transistor for a Biosensor. *Nano Lett.* **2010**, *10*, 2934–2938. [CrossRef]
24. Sarkar, D.; Banerjee, K. Proposal for tunnel-field-effect-transistor as ultra-sensitive and label-free biosensors. *Appl. Phys. Lett.* **2012**, *100*, 143108. [CrossRef]
25. Rigante, S.; Scarbolo, P.; Wipf, M.; Stoop, R.L.; Bedner, K.; Buitrago, E.; Bazigos, A.; Bouvet, D.; Calame, M.; Schönenberger, M.; et al. Sensing with Advanced Computing Technology: Fin Field-Effect Transistors with High-k Gate Stack on Bulk Silicon. *ACS Nano* **2015**, *9*, 4872–4881. [CrossRef] [PubMed]
26. Zhang, Q.; Tu, H.; Yin, H.; Wei, F.; Zhao, H.; Xue, C.; Wei, Q.; Zhang, Z.; Zhang, X.; Zhang, S.; et al. Si Nanowire Biosensors Using a FinFET Fabrication Process for Real Time Monitoring Cellular Ion Actitivies. In Proceedings of the 2018 IEEE International Electron Devices Meeting (IEDM), San Francisco, CA, USA, 1–5 December 2018.
27. Rollo, S.; Rani, D.; Leturcq, R.; Olthuis, W.; Pascual Garcia, C. A high aspect ratio Fin-Ion Sensitive Field Effect Transistor: Compromises towards better electrochemical bio-sensing. *Nano Lett.* **2019**, *19*, 2879–2887. [CrossRef] [PubMed]
28. Ahn, J.-H.; Kim, J.-Y.; Choi, K.; Moon, D.-I.; Kim, C.-H.; Seol, M.-L.; Park, T.J.; Lee, S.Y.; Choi, Y.-K. Nanowire FET Biosensors on a Bulk Silicon Substrate. *IEEE Trans. Electron Devices* **2012**, *59*, 2243–2249. [CrossRef]
29. Chhabra, A.; Kumar, A.; Chaujar, R. Sub-20 nm GaAs Junctionless FinFET for biosensing applications. *Vacuum* **2019**, *160*, 467–471. [CrossRef]
30. Ijjada, S.R.; Mannepalli, C.; Hameed Pasha, M. FinFET Modelling Using TCAD. *Lect. Notes Electr. Eng.* **2018**, *434*, 2018.
31. Mao, S. Graphene Field-Effect Transistor Sensors. In *Graphene Bioelectronics*; Elsevier: Amsterdam, The Netherlands, 2018; pp. 113–132. [CrossRef]
32. Wu, X.; Mu, F.; Wang, Y.; Zhao, H. Graphene and Graphene-Based Nanomaterials for DNA Detection: A Review. *Molecules* **2018**, *23*, 2050. [CrossRef]
33. Zhan, B.; Li, C.; Yang, J.; Jenkins, G.; Huang, W.; Dong, X. Graphene Field-Effect Transistor and Its Application for Electronic Sensing. *Small* **2014**, *10*, 4042–4065. [CrossRef]

34. Star, A.; Tu, E.; Niemann, J.; Gabriel, J.-C.P.; Joiner, C.S.; Valcke, C. Label-free detection of DNA hybridization using carbon nanotube network field-effect transistors. *Proc. Natl. Acad. Sci. USA* **2006**, *103*, 921–926. [CrossRef]
35. Fiorillo, A.S.; Tiriolo, R.; Pullano, S.A. Absorption of Urea into zeolite layer integrated with microelectronic circuits. *IEEE Trans. Nanotechnol.* **2015**, *14*, 2015. [CrossRef]
36. Fiorillo, A.S.; Vinko, J.D.; Accattato, F.; Bianco, M.G.; Critello, C.D.; Pullano, S.A. Zeolite-Based Interfaces for CB Sensors. In *Nanostructured Materials for the Detection of CBRN. NATO Science for Peace and Security Series A: Chemistry and Biology*; Bonča, J., Kruchinin, S., Eds.; Springer: Dordrecht, The Netherlands, 2018.
37. Fiorillo, A.S.; Pullano, S.A.; Tiriolo, R.; Vinko, J.D. Iono-Electronic Interface Based on Innovative Low Temperature Zeolite Coated NMOS (Circuits) for Bio-Nanosensor Manufacture, on Nanomaterials for Security. In *NATO Science for Peace and Security Series A: Chemistry and Biology*; Springer: Dordrecht, The Netherlands, 2016.
38. Fiorillo, A.S. Deposition of zeolite thin layer onto silicon wafers for biomedical use. *IEEE Trans. Nanotechnol.* **2012**, *11*, 654–656. [CrossRef]
39. Heerema, S.J.; Schneider, G.F.; Rozemuller, M.; Vicarelli, L.; Zandbergen, H.W.; Dekker, C. 1/f noise in graphene nanopores. *Nanotechnology* **2015**, *26*. [CrossRef]
40. Cristobal, P.A.; Vilela, P.; Sagheer, A.E.; Cabarcos, E.L.; Brown, T.; Muskens, O.L.; Retama, J.R.; Kanaras, A.G. Highly Sensitive DNA Sensor Based on Upconversion Nanoparticles and Graphene Oxide. *ACS Appl. Mater. Interfaces* **2015**, *7*, 12422–12429. [CrossRef]
41. Lin, L.; Liu, Y.; Tang, L.; Li, J. Electrochemical DNA sensor by the assembly of graphene and DNA-conjugated gold nanoparticles with silver enhancement strategy. *Analyst* **2011**, *136*, 4732–4737. [CrossRef]
42. Patel, H.N.; Carroll, I.; Lopez, R., Jr.; Sankararaman, S.; Etienne, C.; Kodigala, S.R.; Paul, M.R.; Postma, H.W.C. DNA-graphene interactions during translocation through nanogaps. *PLoS ONE* **2017**, *12*. [CrossRef]
43. Merchant, C.A.; Healy, K.; Wanunu, M.; Ray, V.; Peterman, N.; Bartel, J.; Fischbein, M.D.; Venta, K.; Luo, Z.; Johnson, A.T.C.; et al. DNA Translocation through Graphene Nanopores. *Nano Lett.* **2010**, *10*, 2915–2921. [CrossRef]
44. Kim, D.H.; Oh, H.G.; Park, W.H.; Jeon, D.C.; Lim, K.M.; Kim, H.J.; Jang, B.K.; Song, K.S. Detection of Alpha-Fetoprotein in Hepatocellular Carcinoma Patient Plasma with Graphene Field-Effect Transistor. *Sensors* **2018**, *18*, 4032. [CrossRef]
45. Khodagholy, D.; Doublet, T.; Quilichini, P.; Gurfinkel, M.; Leleux, P.; Ghestem, A.; Ismailova, E.; Hervé, T.; Sanaur, S.; Bernard, C.; et al. In vivo recordings of brain activity using organic transistors. *Nat. Commun.* **2013**, *4*, 1–7. [CrossRef]
46. Dimitrakopoulos, C.D.; Malenfant, P.R.L. Organic Thin Film Transistors for Large Area Electronics. *Adv. Mater.* **2002**, *14*, 99–117. [CrossRef]
47. Tarabella, G.; Santato, C.; Yoon Yang, S.; Iannotta, S.; Malliaras, G.G.; Cicoira, F. Effect of the gate electrode on the response of organic electrochemical transistors. *Appl. Phys. Lett.* **2010**, *97*. [CrossRef]
48. Yang, S.Y.; Cicoira, F.; Byrne, R.; Benito-Lopez, F.; Diamond, D.; Owens, R.M.; Malliaras, G.G. Electrochemical transistors with ionic liquids for enzymatic sensing. *Chem. Commun.* **2010**, *46*, 7972–7974. [CrossRef] [PubMed]
49. Lin, P.; Luo, X.T.; Hsing, I.M.; Yan, F. Organic Electrochemical Transistors Integrated in Flexible Microfluidic Systems and Used for Label-Free DNA Sensing. *Adv. Mater.* **2011**, *23*, 4035–4040. [CrossRef] [PubMed]
50. Persson, K.M.; Karlsson, R.; Svennersten, K.; Löffler, S.; Jager, E.W.H.; Richter-Dahlfors, A.; Konradsson, P.; Berggren, M. Electronic control of cell detachment using a self-doped conducting polymer. *Adv. Mater.* **2011**, *23*, 4403–4408. [CrossRef] [PubMed]
51. Tarabella, G.; D'Angelo, P.; Cifarelli, A.; Dimonte, A.; Romeo, A.; Berzina, T.; Erokhina, V.; Iannotta, S. A hybrid living/organic electrochemical transistor based on the Physarum polycephalum cell endowed with both sensing and memristive properties. *Chem. Sci.* **2015**, *6*, 2859–2868. [CrossRef] [PubMed]
52. Gkoupidenis, P.; Koutsouras, D.A.; Malliaras, G.G. Neuromorphic device architectures with global connectivity through electrolyte gating. *Nat. Commun.* **2017**, *8*, 1–8. [CrossRef]
53. Gkoupidenis, P.; Schaefer, N.; Garlan, B.; Malliaras, G.G. Neuromorphic Functions in PEDOT: PSS Organic Electrochemical Transistors. *Adv. Mater.* **2015**, *27*, 7176–7180. [CrossRef]
54. Van De Burgt, Y.; Melianas, A.; Keene, S.T.; Malliaras, G.; Salleo, A. Organic electronics for neuromorphic computing. *Nat. Electron.* **2018**, *1*, 386–397. [CrossRef]

55. Stavrinidou, E.; Gabrielsson, R.; Gomez, E.; Crispin, X.; Nilsson, O.; Simon, D.T.; Berggren, M. Electronic plants. *Sci. Adv.* **2015**, *1*. [CrossRef]
56. Yeung, S.Y.; Gu, X.; Tasng, C.M.; Tsao, S.W.; Hsing, I.-M. Engineering organic electrochemical transistor (OECT) to be sensitive cell-based biosensor through tuning of channel area. *Sens. Actuators A Phys.* **2019**, *287*, 185–193. [CrossRef]
57. Zhu, Z.T.; Mabeck, J.T.; Zhu, C.; Cady, N.C.; Batt, C.A.; Malliaras, G.G. A simple poly (3,4-ethylene dioxythiophene)/poly(styrene sulfonic acid) transistor for glucose sensing at neutral pH. *Chem. Commun.* **2004**, 1556–1557. [CrossRef]
58. Lin, P.; Yan, F.; Chan, H.L.W. Ion-Sensitive Properties of Organic Electrochemical Transistors. *ACS Appl. Mater. Interfaces* **2010**, *2*, 1637–1641. [CrossRef] [PubMed]
59. Kim, S.-M.; Kim, C.-H.; Kim, Y.; Kim, N.; Lee, W.-J.; Lee, E.-H.; Kim, D.; Park, S.; Lee, K.; Rivnay, J.; et al. Influence of PEDOT:PSS crystallinity and composition on electrochemical transistor performance and long-term stability. *Nat. Commun.* **2018**, *9*. [CrossRef] [PubMed]
60. Spyropoulos, G.D.; Gelinas, J.N.; Khodagholy, D. Internal ion-gated organic electrochemical transistor: A building block for integrated bioelectronics. *Sci. Adv.* **2019**, *5*. [CrossRef] [PubMed]
61. Hempel, F.; Law, J.; Susloparova, A.; Pachauri, V.; Ingebrandt, S. PEDOT: PSS Organic Electrochemical Transistors for Cell Sensing Applications. *Front. Neurosci.* **2016**, *10*. [CrossRef]
62. Bai, L.; García Elósegui, C.; Li, W.; Yu, P.; Fei, J.; Mao, L. Biological Applications of Organic Electrochemical Transistors: Electrochemical Biosensors and Electrophysiology Recording. *Front. Chem.* **2019**, *7*, 313. [CrossRef]
63. Macaya, D.J.; Nikolou, M.; Takamatsu, S.; Mabeck, J.T.; Owens, R.M.; Malliaras, G.G. Simple glucose sensors with micromolar sensitivity based on organic electrochemical transistors. *Sens. Actuator B Chem.* **2007**, *123*, 374–378. [CrossRef]
64. Hamedi, M.; Forchheimer, R.; Inganas, O. Towards woven logic from organic electronic fibres. *Nat. Mater.* **2007**, *6*, 357–362. [CrossRef]
65. Bernards, D.A.; Macaya, D.J.; Nikolou, M.; DeFranco, J.A.; Takamatsu, S.; Malliaras, G.G. Enzymatic sensing with organic electrochemical transistors. *J. Mater. Chem.* **2008**, *18*, 116. [CrossRef]
66. Lin, P.; Yan, F.; Yu, J.; Chan, H.L.W.; Yang, M. The application of organic electrochemical transistors in cell-based biosensors. *Adv. Mater.* **2010**, *22*, 3655–3660. [CrossRef]
67. Kim, D.J.; Lee, N.E.; Park, J.S.; Park, I.J.; Kim, J.G.; Cho, H.J. Organic electrochemical transistor based immunosensor for prostate specific antigen (PSA) detection using gold nanoparticles for signal amplification. *Biosens. Bioelectron.* **2010**, *25*, 2477–2482. [CrossRef]
68. Coppedè, N.; Villani, M.; Gentile, F. Diffusion Driven Selectivity in Organic Electrochemical Transistors. *Sci. Rep.* **2014**, *4*, 1–7. [CrossRef] [PubMed]
69. Gentile, F.; Ferrara, L.; Villani, M.; Bettelli, M.; Iannotta, S.; Zappettini, A.; Cesarelli, M.; Di Fabrizio, E.; Coppedè, N. Geometrical Patterning of Super-Hydrophobic Biosensing Transistors Enables Space and Time Resolved Analysis of Biological Mixtures. *Sci. Rep.* **2016**, *6*, 1–15. [CrossRef] [PubMed]
70. Tam, D.; von Arnim, V.; McKinley, G.H.; Hosoi, A.E. Marangoni convection in droplets on superhydrophobic surfaces. *J. Fluid Mech.* **2009**, *624*, 101–123. [CrossRef]
71. Astier, Y.; Data, L.; Carney, R.P.; Stellacci, F.; Gentile, F.; Di Fabrizio, E. Artificial Surface-Modified Si3N4 Nanopores for Single Surface-Modified Gold Nanoparticle Scanning. *Small* **2010**, *7*, 455–459. [CrossRef] [PubMed]
72. Kim, M.-M.; Zydney, A.L. Effect of electrostatic, hydrodynamic, and Brownian forces on particle trajectories and sieving in normal flow filtration. *J. Colloid Interface Sci.* **2004**, *269*, 425–431. [CrossRef] [PubMed]
73. Gentile, F.; Coluccio, M.L.; Zaccaria, R.P.; Francardi, M.; Cojoc, G.; Perozziello, G.; Raimondo, R.; Candeloroa, P.; Fabrizio, E.D. Selective on site separation and detection of molecules in diluted solutions with superhydrophobic clusters of plasmonic nanoparticles. *Nanoscale* **2014**, *6*, 8208–8225. [CrossRef] [PubMed]
74. Lafuma, A.; Quéré, D. Superhydrophobic states. *Nat. Mater.* **2003**, *2*, 457–460. [CrossRef]
75. Ueda, E.; Levkin, P.A. Emerging Applications of Superhydrophilic-Superhydrophobic Micropatterns. *Adv. Mater.* **2013**, *25*, 1234–1247. [CrossRef]
76. Hoshian, S.; Jokinen, V.; Somerkivi, V.; Lokanathan, A.R.; Franssila, S. Robust Superhydrophobic Silicon without a Low Surface-Energy Hydrophobic Coating. *ACS Appl. Mater. Interfaces* **2015**, *7*, 941–949. [CrossRef]

77. He, Y.; Jiang, C.; Yin, H.; Yuan, W. Tailoring the wettability of patterned silicon surfaces with dual-scale pillars: From hydrophilicity to superhydrophobicity. *Appl. Surf. Sci.* **2011**, *257*, 7689–7692. [CrossRef]
78. Xiu, Y.; Zhu, L.; Dennis, W.; Hess, C.; Wong, P. Hierarchical Silicon Etched Structures for Controlled Hydrophobicity/Superhydrophobicity. *Nano Lett.* **2007**, *7*, 3388–3393. [CrossRef] [PubMed]
79. Falde, E.J.; Yohe, S.T.; Colson, Y.L.; Grinstaff, M.W. Superhydrophobic materials for biomedical applications. *Biomaterials* **2016**, *104*, 87–103. [CrossRef] [PubMed]
80. Gentile, F.; Battista, E.; Accardo, A.; Coluccio, M.; Asande, M.; Perozziello, G.; Das, G.; Liberale, C.; De Angelis, F.; Candeloro, P.; et al. Fractal Structure Can Explain the Increased Hydrophobicity of NanoPorous Silicon Films. *Microelectron. Eng.* **2011**, *88*, 2537–2540. [CrossRef]
81. Bartolo, D.; Bouamrirene, F.; Verneuil, É.; Buguin, A.; Silberzan, P.; Moulinet, S. Bouncing or sticky droplets: Impalement transitions on superhydrophobic micropatterned surfaces. *Europhys. Lett.* **2006**, *74*, 2. [CrossRef]
82. Gentile, F.; Coppedè, N.; Tarabella, G.; Villani, M.; Calestani, D.; Candeloro, P.; Iannotta, S.; Di Fabrizio, E. Microtexturing of the Conductive PEDOT:PSS Polymer for Superhydrophobic Organic Electrochemical Transistors. *BioMed. Res. Int.* **2014**, 1–10. [CrossRef]
83. Malara, N.; Gentile, F.; Coppedè, N.; Coluccio, M.L.; Candeloro, P.; Perozziello, G.; Ferrara, L.; Giannetto, M.; Careri, M.; Castellini, A.; et al. Superhydrophobic lab-on-chip measures secretome protonation state and provides a personalized risk assessment of sporadic tumour. *NPJ Precis. Oncol.* **2018**, *2*. [CrossRef]
84. Farzin, L.; Shamsipur, M. Recent advances in design of electrochemical affinity biosensors for low level detection of cancer protein biomarkers using nanomaterial-assisted signal enhancement strategies. *J. Pharmac. Biomed. Anal.* **2018**, *147*, 185–210. [CrossRef]
85. Lim, S.A.; Ahmed, A.U. Electrochemical immunosensors and their recent nanomaterial-based signal amplification strategies: A review. *RSC Adv.* **2016**, *6*, 24995–25014. [CrossRef]
86. Xia, Y.Y.; Si, J.; Li, Z.Y. Fabrication techniques for microfluidic paper-based analytical devices and their applications for biological testing: A review. *Biosens. Bioelectron.* **2016**, *77*, 774–789. [CrossRef]
87. Balu, B.; Berry, A.D.; Hess, D.W.; Breedveld, V. Patterning of superhydrophobic paper to control the mobility of micro-liter drops for two-dimensional lab-on-paper applications. *Lab Chip* **2009**, *9*, 3066–3075. [CrossRef]
88. Kauffman, P.; Fu, E.; Lutz, B.; Yager, P. Visualization and measurement of flow in two-dimensional paper networks. *Lab Chip* **2010**, *10*, 2614–2617. [CrossRef] [PubMed]
89. Kalisha, B.; Tsutsui, H. Patterned adhesive enables construction of nonplanar three-dimensional paper microfluidic circuits. *Lab Chip* **2014**, *14*, 4354–4361. [CrossRef] [PubMed]
90. Mosadegh, B.; Dabiri, B.E.; Lockett, M.R.; Derda, R.; Campbell, P.; Parker, K.K.; Whitesides, G.M. Three-dimensional paper-based model for cardiac ischemia. *Adv. Healthc. Mater.* **2014**, *3*, 1036–1043. [CrossRef]
91. Li, X.; Tian, J.; Garnier, G.; Shen, W. Fabrication of paper-based microfluidic sensors by printing. *Surf. B* **2010**, *76*, 564–570. [CrossRef]
92. He, Y.; Wu, Y.; Xiao, X.; Fu, J.Z.; Xue, G.H. A low-cost and rapid microfluidic paper-based analytical device fabrication method: Flash foam stamp lithography. *RSC Adv.* **2014**, *4*, 63860–63865. [CrossRef]
93. Olkkonen, J.; Lehtinen, K.; Erho, T. Flexographically Printed Fluidic Structures in Paper. *Anal. Chem.* **2010**, *82*, 10246–10250. [CrossRef]
94. Yan, C.F.; Yu, S.Y.; Jiang, Y.; He, Q.H.; Chen, H.W. Fabrication of Paper-based Microfluidic Devices by Plasma Treatment and Its Application in Glucose Determination. *Acta Chim. Sin.* **2014**, *72*, 1099–1104. [CrossRef]
95. Sones, C.L.; Katis, I.N.; He, P.J.W.; Mills, B.; Namiq, M.F.; Shardlow, P.; Ibsen, M.; Eason, R.W. Laser-induced photo-polymerisation for creation of paper-based fluidic devices. *Lab Chip* **2014**, *14*, 4567–4574. [CrossRef]
96. Cai, L.; Xu, C.; Lin, S.; Luo, J.; Wu, M.; Yang, F. A simple paper-based sensor fabricated by selective wet etching of silanized filter paper using a paper mask. *Biomicrofluidics* **2014**, *8*, 5. [CrossRef]
97. Renault, C.; Scida, K.; Knust, K.N.; Fosdick, S.E.; Crooks, R.M. Paper-Based Bipolar Electrochemistry. *J. Electrochem. Sci. Tech.* **2013**, *4*, 146–152. [CrossRef]
98. Dungchai, W.; Chailapakul, O.; Henry Analyst, C.S. A low-cost, simple, and rapid fabrication method for paper-based microfluidics using wax screen-printing. *Analyst* **2011**, *136*, 77–82. [CrossRef]
99. Liang, L.L.; Su, M.; Li, L.; Lan, F.F.; Yang, G.X.; Ge, S.G.; Yu, J.H.; Song, X.R. Aptamer-based fluorescent and visual biosensor for multiplexed monitoring of cancer cells in microfluidic paper-based analytical devices. *Sens. Actuator B Chem.* **2016**, *229*, 347–354. [CrossRef]

100. Rattanarat, P.; Dungchai, W.; Cate, D.; Volckens, J.; Chailapakul, O.; Henry, C.S. Multilayer paper-based device for colorimetric and electrochemical quantification of metals. *Anal. Chem.* **2014**, *86*, 3555–3562. [CrossRef] [PubMed]
101. Lopez-Ruiz, N.; Curto, V.F.; Erenas, M.M.; Benito-Lopez, F.; Diamond, D.; Palma, A.J.; Capitan-Vallvey, L.F. Smartphone-based simultaneous pH and nitrite colorimetric determination for paper microfluidic devices. *Anal. Chem.* **2014**, *86*, 9554–9562. [CrossRef] [PubMed]
102. Bonapace, G.; Marasco, O.; Scozzafava, G.; Michael, A.; Pittelli, M.; Greto, T.; Moricca, M.T.; Vismara, S.A.; Valentini, A.; Vismara, M.F.M.; et al. GP15 Phenylalanine measurements in human blood using NIR spectroscopy and DBS, a preliminary study. *Arch. Dis. Child.* **2019**, *104*, A35–A36.
103. Moral-Munoz, J.A.; Zhang, W.; Cobo, M.J.; Herrera-Viedma, E.; Kaber, D.B. Smartphone-based systems for physical rehabilitation applications: A systematic review. *Assist. Technol.* **2019**, 1–14. [CrossRef]
104. Nussbaum, Y.; Kelly, C.; Quinby, E.; Mac, A.; Parmanto, B.; Dicianno, B.E. Systematic Review of Mobile Health Applications in Rehabilitation. *Arch. Phys. Med. Rehabil.* **2019**, *100*, 115–127. [CrossRef]
105. Ruzycka, M.; Cimpan, M.R.; Rios-Mondragon, I.; Grudzinski, I.P. Microfluidics for studying metastatic patterns of lung cancer. *J. Nanobiotechnol.* **2019**, *17*, 71. [CrossRef]
106. Mohammadniaei, M.; Nguyen, H.V.; Tieu, M.V.; Lee, M.H. 2D Materials in Development of Electrochemical Point-of-Care Cancer Screening Devices. *Micromachines (Basel)* **2019**, *10*, 662. [CrossRef]
107. Mohammadniaei, M.; Park, C.; Min, J.; Sohn, H.; Lee, T. Fabrication of Electrochemical-Based Bioelectronic Device and Biosensor Composed of Biomaterial-Nanomaterial Hybrid. In *Biomimetic Medical Materials: From Nanotechnology to 3D Bioprinting*; Noh, I., Ed.; Springer: Singapore, 2018; pp. 263–296.
108. Sajid, M.; Kawde, A.-N.; Daud, M. Designs, formats and applications of lateral flow assay: A literature review. *J. Saudi Chem. Soc.* **2015**, *19*, 689–705. [CrossRef]
109. Sher, M.; Zhuang, R.; Demirci, U.; Asghar, W. Paper-based analytical devices for clinical diagnosis: Recent advances in the fabrication techniques and sensing mechanisms. *Expert. Rev. Mol. Diagn.* **2017**, *17*, 351–366. [CrossRef] [PubMed]
110. Ali, M.H.; Elsherbiny, M.E.; Emara, M. Updates on Aptamer Research. *Int. J. Mol. Sci.* **2019**, *20*, 2511. [CrossRef] [PubMed]
111. Bukari, B.A.; Citartan, M.; Ch'ng, E.S.; Bilibana, M.P.; Rozhdestvensky, T.; Tang, T.H. Aptahistochemistry in diagnostic pathology: Technical scrutiny and feasibility. *Histochem. Cell Biol.* **2017**, *147*, 545–553. [CrossRef] [PubMed]
112. Huh, H.Y.; Kim, S.; Lee, D. Isolation of rna aptamers targeting HER-2-overexpressing breast cancer cells using cell-SELEX. *Bull. Korean Chem. Soc.* **2009**, *30*, 1827–1831. [CrossRef]
113. Niazi, J.; Verma, S.; Niazi, S.; Qureshi, A. In vitro HER2 protein-induced affinity dissociation of carbon nanotube-wrapped anti-HER2 aptamers for HER2 protein detection. *Analyst* **2013**, *140*, 243–249. [CrossRef]
114. Ahirwar, R.; Vellarikkal, S.K.; Sett, A.; Sivasubbu, S.; Scaria, V.; Bora, U.; Borthakur, B.B.; Kataki, A.C.; Sharma, J.D.; Nahar, P. Aptamer-assisted detection of the altered expression of estrogen receptor alpha in human breast cancer. *PLoS ONE* **2016**, *11*. [CrossRef]
115. Lan, J.; Li, L.; Liu, Y.; Yan, L.; Li, C.; Chen, J.; Chen, X. Upconversion luminescence assay for the detection of the vascular endothelial growth factor, a biomarker for breast cancer. *Microchim. Acta.* **2016**, *183*, 3201–3208. [CrossRef]
116. Meirinho, S.G.; Dias, L.G.; Peres, A.M.; Rodrigues, L.R. Development of an electrochemical aptasensor for the detection of human osteopontin. *Procedia Eng.* **2014**, *87*, 316–319. [CrossRef]
117. Hua, X.; Zhou, Z.; Yuan, L.; Liu, S. Selective collection and detection of MCF-7 breast cancer cells using aptamer-functionalized magnetic beads and quantum dots based nano-bio-probes. *Anal. Chim. Acta.* **2013**, *788*, 135–140. [CrossRef]
118. Wu, J.; Wang, C.; Li, X.; Song, Y.; Wang, W.; Li, C.; Hu, J.; Zhu, Z.; Li, J.; Zhang, W.; et al. Identification, characterization and application of a G-quadruplex structured DNA aptamer against cancer biomarker protein anterior gradient homolog 2. *PLoS ONE* **2012**, *7*, e46393. [CrossRef]

119. Roointan, A.; Ahmad Mir, T.; Ibrahim Wani, S.; Mati-Ur-Rehman Hussain, K.K.; Ahmed, B.; Abrahim, S.; Savardashtaki, A.; Gandomani, G.; Gandomani, M.; Chinnappan, R.; et al. Early detection of lung cancer biomarkers through biosensor technology: A review. *J. Pharm. Biomed. Anal.* **2019**, *164*, 93–103. [CrossRef] [PubMed]
120. Ye, H.; Qin, B.; Sun, Y.; Li, J. Electrochemical Detection of VEGF165 Lung Cancer Marker Based on Au-Pd Alloy Assisted Aptasenor. *Int. J. Electrochem. Sci.* **2017**, *12*, 1818–1828. [CrossRef]
121. Tabrizi, M.A.; Shamsipur, M.; Farzin, L. A high sensitive electrochemical aptasensor for the determination of VEGF 165 in serum of lung cancer patient. *Biosens. Bioelectron.* **2015**, *74*, 764–769. [CrossRef] [PubMed]
122. Xu, X.-W.; Weng, X.-H.; Wang, C.-L.; Lin, W.-W.; Liu, A.-L.; Chen, W.; Lin, X.-H. Detection EGFR exon 19 status of lung cancer patients by DNA electrochemical biosensor. *Biosens. Bioelectron.* **2016**, *80*, 411–417. [CrossRef] [PubMed]
123. Kim, D.-M.; Noh, H.-B.; Park, D.S.; Ryu, S.-H.; Koo, J.S.; Shim, Y.-B. Immunosensors for detection of Annexin II and MUC5AC for early diagnosis of lung cancer. *Biosens. Bioelectron* **2009**, *25*, 456–462. [CrossRef]
124. Hussain, K.K.; Gurudatt, N.; Mir, T.A.; Shim, Y.-B. Amperometric sensing of HIF1α expressed in cancer cells and the effect of hypoxic mimicking agents. *Biosens. Bioelectron* **2016**, *83*, 312–318. [CrossRef]
125. Akhtar, M.H.; Mir, T.A.; Gurudatt, N.; Chung, S.; Shim, Y.-B. Sensitive NADH detection in a tumorigenic cell line using a nano-biosensor based on the organic complex formation. *Biosens. Bioelectron* **2016**, *85*, 488–495. [CrossRef]
126. Kościelniak-Merak, B.; Radosavljević, B.; Zając, A.; Tomasik, P.J. Faecal Occult Blood Point-of-Care Tests. *J. Gastrointest. Cancer* **2018**, *49*, 402–405. [CrossRef]
127. Huddy, J.R.; Ni, M.Z.; Markar, S.R.; Hanna, G.B. Point-of-care testing in the diagnosis of gastrointestinal cancers: Current technology and future directions. *World J. Gastroenterol.* **2015**, *21*, 4111–4120. [CrossRef]
128. Tinmouth, J.; Lansdorp-Vogelaar, I.; Allison, J.E. Faecal immunochemical tests versus guaiac faecal occult blood tests: What clinicians and colorectal cancer screening programme organisers need to know. *Gut* **2015**, *64*, 1327–1337. [CrossRef]
129. Faivre, J.; Dancourt, V.; Lejeune, C. Screening for colorectal cancer with immunochemical faecal occult blood tests. *Dig. Liver Dis.* **2012**, *44*, 967–973. [CrossRef] [PubMed]
130. Huang, Y.; Li, Q.; Ge, W.; Cai, S.; Zhang, S.; Zheng, S. Predictive power of quantitative and qualitative fecal immunochemical tests for hemoglobin in population screening for colorectal neoplasm. *Eur. J. Cancer Prev.* **2014**, *23*, 27–34. [CrossRef] [PubMed]
131. Leontiadis, G.I. Fecal immunochemical tests in patients at increased risk for colorectal cancer-is it prime time yet? *JAMA Intern. Med.* **2017**, *177*, 1119–1120. [CrossRef] [PubMed]
132. Imperiale, T.; Ransohoff, D.F.; Itzkowitz, S.H.; Turnbull, B.A.; Ross, M.E.; Colorectal Cancer Study Group. Fecal DNA versus fecal occult blood for colorectal cancer screening in an average-risk population. *N. Engl. J. Med.* **2004**, *351*, 2704–2714. [CrossRef] [PubMed]
133. Imperiale, T.F.; Ransohoff, D.F.; Itzkowitz, S.H.; Levin, T.R.; Lavin, P.; Lidgard, G.P.; Ahlquist, D.A.; Berger, B.M. Multitarget stool DNA testing for colorectal-cancer screening. *N. Engl. J. Med.* **2014**, *370*, 1287–1297. [CrossRef] [PubMed]
134. Ahlquist, D.A.; Zou, H.; Domanico, M.; Mahoney, D.W.; Yab, T.C.; Taylor, W.R.; Butz, M.L.; Thibodeau, S.N.; Rabeneck, L.; Paszat, L.F.; et al. Next-generation stool DNA test accurately detects colorectal cancer and large adenomas. *Gastroenterology* **2012**, *142*, 248–256. [CrossRef] [PubMed]
135. Mignogna, C.; Staropoli, N.; Botta, C.; De Marco, C.; Rizzuto, A.; Morelli, M.; Di Cello, A.; Franco, R.; Camastra, C.; Presta, I.; et al. Aurora Kinase A expression predicts platinum-resistance and adverse outcome in high-grade serous ovarian carcinoma patients. *J. Ovarian Res.* **2016**, *9*, 31. [CrossRef]
136. Frossard, J.L.; Peyer, R. Fecal DNA for colorectal cancer screening. *N. Engl. J. Med.* **2005**, *352*, 1384–1385.
137. Church, T.R.; Wandell, M.; Lofton-Day, C.; Mongin, S.J.; Burger, M.; Payne, S.R.; Castaños-Vélez, E.; Blumenstein, B.A.; Rösch, T.; Osborn, N.; et al. Prospective evaluation of methylated SEPT9 in plasma for detection of asymptomatic colorectal cancer. *Gut* **2014**, *63*, 317–325. [CrossRef]
138. Ren, A.; Dong, Y.; Tsoi, H.; Yu, J. Detection of miRNA as non-invasive biomarkers of colorectal cancer. *Int. J. Mol. Sci.* **2015**, *16*, 2810–2823. [CrossRef]
139. Mark, M.; Phillip, N.; Alison, H.; Sean, C.; Sarah, F.; Logan, J.; Joy, C.; Shane, M.; Richard, S.; Citorik Robert, J.; et al. An ingestible bacterial-electronic system to monitor gastrointestinal health. *Science* **2018**, *360*, 915–918. [CrossRef]

140. Pundir, C.S.; Deswal, R.; Yadav, P. Quantitative analysis of sarcosine with special emphasis on biosensors: A review. *Biomarkers* **2019**, 1–30. [CrossRef] [PubMed]
141. Kozel, T.R.; Burnham-Marusich, A.R. Point-of-Care Testing for Infectious Diseases: Past, Present, and Future. *J. Clin. Microbiol.* **2017**, *55*, 2313–2320. [CrossRef]
142. Zanoli, L.M.; Spoto, G. Isothermal Amplification Methods for the Detection of Nucleicl Acids in Microfluidic Devices. *Biosensors* **2013**, *3*, 18–43. [CrossRef] [PubMed]
143. Bell, J.; Bonner, A.; Cohen, D.M.; Birkhahn, R.; Yogev, R.; Triner, W.; Cohen, J.; Palavecino, E.; Selvarangan, R. Multicenter clinical evaluation of the novel Alere i Influenza A&B isothermal nucleic acid amplification test. *J. Clin. Virol.* **2014**, *61*, 81–86. [PubMed]
144. Cohen, D.M.; Russo, M.E.; Jaggi, P.; Kline, J.; Gluckman, W.; Parekh, A. Multicenter clinical evaluation of the novel Alere i Strep A isothermal nucleic acid amplification test. *J. Clin. Microbiol.* **2015**, *53*, 2258–2261. [CrossRef]
145. Hassan, F.; Hays, L.M.; Bonner, A.; Bradford, B.J.; Franklin, R., Jr.; Hendry, P.; Kaminetsky, J.; Vaughn, M.; Cieslak, K.; Moffatt, M.E.; et al. Multicenter Clinical Evaluation of the Alere i Respiratory Syncytial Virus Isothermal Nucleic Acid Amplification Assay. *J. Clin. Microbiol.* **2018**, *56*. [CrossRef]
146. Grebely, J.; Applegate, T.L.; Cunningham, P.; Feld, J.J. Hepatitis C point-of-care diagnostics: In search of a single visit diagnosis. *Expert Rev. Mol. Diagn.* **2017**, *17*, 1109–1115. [CrossRef]
147. Gaydos, C.A.; Klausner, J.D.; Pai, N.P.; Kelly, H.; Coltart, C.; Peeling, R.W. Rapid and point-of-care tests for the diagnosis of Trichomonas vaginalis in women and men. *Sex. Transm. Infect.* **2017**, *93*, S31–S35. [CrossRef]
148. Hurt, C.B.; Nelson, J.A.E.; Hightow-Weidman, L.B.; Miller, W.C. Selecting an HIV Test: A Narrative Review for Clinicians and Researchers. *Sex. Transm. Dis.* **2017**, *44*, 739–746. [CrossRef]
149. Kelly, H.; Coltart, C.E.M.; Pant Pai, N.; Klausner, J.D.; Unemo, M.; Toskin, I.; Peeling, R.W. Systematic reviews of point-of-care tests for the diagnosis of urogenital Chlamydia trachomatis infections. *Sex. Transm. Infect.* **2017**, *93*, S22–S30. [CrossRef] [PubMed]
150. Basile, K.; Kok, J.; Dwyer, D.E. Point-of-care diagnostics for respiratory viral infections. *Expert Rev. Mol. Diagn.* **2018**, *18*, 75–83. [CrossRef] [PubMed]
151. Nzelu, C.O.; Kato, H.; Peters, N.C. Loop-mediated isothermal amplification (LAMP): An advanced molecular point-of-care technique for the detection of Leishmania infection. *PLoS Negl. Trop. Dis.* **2019**, *13*, e0007698. [CrossRef] [PubMed]
152. Datta, M.; Desai, D.; Kumar, A. Gene Specific DNA Sensors for Diagnosis of Pathogenic Infections. *Indian J. Microbiol.* **2017**, *57*, 139–147. [CrossRef] [PubMed]
153. Syal, K.; Mo, M.; Yu, H.; Iriya, R.; Jing, W.; Guodong, S.; Wang, S.; Grys, T.E.; Haydel, S.E.; Tao, N. Current and emerging techniques for antibiotic susceptibility tests. *Theranostics* **2017**, *10*, 1795–1805. [CrossRef] [PubMed]
154. Maugeri, G.; Lychko, I.; Sobral, R.; Roque, A.C.A. Identification and Antibiotic-Susceptibility Profiling of Infectious Bacterial Agents: A Review of Current and Future Trends. *Biotechnol. J.* **2019**, *14*, 1700750. [CrossRef]
155. Morris, E.; McCartney, D.; Lasserson, D.; Van den Bruel, A.; Fisher, R.; Hayward, G. Point-of-care lactate testing for sepsis at presentation to health care: A systematic review of patient outcomes. *Br. J. Gen. Pract.* **2017**, *67*, e859–e870. [CrossRef]
156. Baig, M.A.; Shahzad, H.; Hussain, E.; Mian, A. Validating a point of care lactate meter in adult patients with sepsis presenting to the emergency department of a tertiary care hospital of a low- to middle-income country. *World J. Emerg. Med.* **2017**, *8*, 184–189. [CrossRef]
157. Hassan, U.; Ghonge, T.; Reddy, B., Jr.; Patel, M.; Rappleye, M.; Taneja, I.; Tanna, A.; Healey, R.; Manusry, N.; Price, Z.; et al. A point-of-care microfluidic biochip for quantification of CD64 expression from whole blood for sepsis stratification. *Nat. Commun.* **2017**, *8*, 1–12. [CrossRef]
158. Westra, B.L.; Landman, S.; Yadav, P.; Steinbach, M. Secondary Analysis of an Electronic Surveillance System Combined with Multi-focal Interventions for Early Detection of Sepsis. *Appl. Clin. Inform.* **2017**, *8*, 47–66. [CrossRef]

MDPI
St. Alban-Anlage 66
4052 Basel
Switzerland
Tel. +41 61 683 77 34
Fax +41 61 302 89 18
www.mdpi.com

Micromachines Editorial Office
E-mail: micromachines@mdpi.com
www.mdpi.com/journal/micromachines

www.ingramcontent.com/pod-product-compliance
Lightning Source LLC
LaVergne TN
LVHW070631170726
843515LV00004B/223
9783036515700